SEISMIC HAZARDS IN
SITE EVALUATION FOR
NUCLEAR INSTALLATIONS

The following States are Members of the International Atomic Energy Agency:

AFGHANISTAN
ALBANIA
ALGERIA
ANGOLA
ANTIGUA AND BARBUDA
ARGENTINA
ARMENIA
AUSTRALIA
AUSTRIA
AZERBAIJAN
BAHAMAS
BAHRAIN
BANGLADESH
BARBADOS
BELARUS
BELGIUM
BELIZE
BENIN
BOLIVIA, PLURINATIONAL
   STATE OF
BOSNIA AND HERZEGOVINA
BOTSWANA
BRAZIL
BRUNEI DARUSSALAM
BULGARIA
BURKINA FASO
BURUNDI
CAMBODIA
CAMEROON
CANADA
CENTRAL AFRICAN
   REPUBLIC
CHAD
CHILE
CHINA
COLOMBIA
COMOROS
CONGO
COSTA RICA
CÔTE D'IVOIRE
CROATIA
CUBA
CYPRUS
CZECH REPUBLIC
DEMOCRATIC REPUBLIC
   OF THE CONGO
DENMARK
DJIBOUTI
DOMINICA
DOMINICAN REPUBLIC
ECUADOR
EGYPT
EL SALVADOR
ERITREA
ESTONIA
ESWATINI
ETHIOPIA
FIJI
FINLAND
FRANCE
GABON

GEORGIA
GERMANY
GHANA
GREECE
GRENADA
GUATEMALA
GUYANA
HAITI
HOLY SEE
HONDURAS
HUNGARY
ICELAND
INDIA
INDONESIA
IRAN, ISLAMIC REPUBLIC OF
IRAQ
IRELAND
ISRAEL
ITALY
JAMAICA
JAPAN
JORDAN
KAZAKHSTAN
KENYA
KOREA, REPUBLIC OF
KUWAIT
KYRGYZSTAN
LAO PEOPLE'S DEMOCRATIC
   REPUBLIC
LATVIA
LEBANON
LESOTHO
LIBERIA
LIBYA
LIECHTENSTEIN
LITHUANIA
LUXEMBOURG
MADAGASCAR
MALAWI
MALAYSIA
MALI
MALTA
MARSHALL ISLANDS
MAURITANIA
MAURITIUS
MEXICO
MONACO
MONGOLIA
MONTENEGRO
MOROCCO
MOZAMBIQUE
MYANMAR
NAMIBIA
NEPAL
NETHERLANDS
NEW ZEALAND
NICARAGUA
NIGER
NIGERIA
NORTH MACEDONIA
NORWAY

OMAN
PAKISTAN
PALAU
PANAMA
PAPUA NEW GUINEA
PARAGUAY
PERU
PHILIPPINES
POLAND
PORTUGAL
QATAR
REPUBLIC OF MOLDOVA
ROMANIA
RUSSIAN FEDERATION
RWANDA
SAINT LUCIA
SAINT VINCENT AND
   THE GRENADINES
SAMOA
SAN MARINO
SAUDI ARABIA
SENEGAL
SERBIA
SEYCHELLES
SIERRA LEONE
SINGAPORE
SLOVAKIA
SLOVENIA
SOUTH AFRICA
SPAIN
SRI LANKA
SUDAN
SWEDEN
SWITZERLAND
SYRIAN ARAB REPUBLIC
TAJIKISTAN
THAILAND
TOGO
TRINIDAD AND TOBAGO
TUNISIA
TURKEY
TURKMENISTAN
UGANDA
UKRAINE
UNITED ARAB EMIRATES
UNITED KINGDOM OF
   GREAT BRITAIN AND
   NORTHERN IRELAND
UNITED REPUBLIC
   OF TANZANIA
UNITED STATES OF AMERICA
URUGUAY
UZBEKISTAN
VANUATU
VENEZUELA, BOLIVARIAN
   REPUBLIC OF
VIET NAM
YEMEN
ZAMBIA
ZIMBABWE

The Agency's Statute was approved on 23 October 1956 by the Conference on the Statute of the IAEA held at United Nations Headquarters, New York; it entered into force on 29 July 1957. The Headquarters of the Agency are situated in Vienna. Its principal objective is "to accelerate and enlarge the contribution of atomic energy to peace, health and prosperity throughout the world".

IAEA SAFETY STANDARDS SERIES No. SSG-9 (Rev. 1)

SEISMIC HAZARDS IN SITE EVALUATION FOR NUCLEAR INSTALLATIONS

SPECIFIC SAFETY GUIDE

INTERNATIONAL ATOMIC ENERGY AGENCY
VIENNA, 2022

COPYRIGHT NOTICE

All IAEA scientific and technical publications are protected by the terms of the Universal Copyright Convention as adopted in 1952 (Berne) and as revised in 1972 (Paris). The copyright has since been extended by the World Intellectual Property Organization (Geneva) to include electronic and virtual intellectual property. Permission to use whole or parts of texts contained in IAEA publications in printed or electronic form must be obtained and is usually subject to royalty agreements. Proposals for non-commercial reproductions and translations are welcomed and considered on a case-by-case basis. Enquiries should be addressed to the IAEA Publishing Section at:

Marketing and Sales Unit, Publishing Section
International Atomic Energy Agency
Vienna International Centre
PO Box 100
1400 Vienna, Austria
fax: +43 1 26007 22529
tel.: +43 1 2600 22417
email: sales.publications@iaea.org
www.iaea.org/publications

© IAEA, 2022

Printed by the IAEA in Austria
January 2022
STI/PUB/1950

IAEA Library Cataloguing in Publication Data

Names: International Atomic Energy Agency.
Title: Seismic hazards in site evaluation for nuclear installations / International Atomic Energy Agency.
Description: Vienna : International Atomic Energy Agency, 2022. | Series: IAEA safety standards series, ISSN 1020–525X ; no. SSG-9 (Rev. 1) | Includes bibliographical references.
Identifiers: IAEAL 21-01454 | ISBN 978-92-0-117821-3 (paperback : alk. paper) | ISBN 978-92-0-117921-0 (pdf) | ISBN 978-92-0-118021-6 (epub)
Subjects: LCSH: Nuclear facilities — Seismic prospecting — Risk assessment. | Nuclear facilities — Seismic tomography — Safety measures. | Earthquake hazard analysis. | Earthquake resistant design.
Classification: UDC 621.039.58 | STI/PUB/1950

FOREWORD

by Rafael Mariano Grossi
Director General

The IAEA's Statute authorizes it to "establish...standards of safety for protection of health and minimization of danger to life and property". These are standards that the IAEA must apply to its own operations, and that States can apply through their national regulations.

The IAEA started its safety standards programme in 1958 and there have been many developments since. As Director General, I am committed to ensuring that the IAEA maintains and improves upon this integrated, comprehensive and consistent set of up to date, user friendly and fit for purpose safety standards of high quality. Their proper application in the use of nuclear science and technology should offer a high level of protection for people and the environment across the world and provide the confidence necessary to allow for the ongoing use of nuclear technology for the benefit of all.

Safety is a national responsibility underpinned by a number of international conventions. The IAEA safety standards form a basis for these legal instruments and serve as a global reference to help parties meet their obligations. While safety standards are not legally binding on Member States, they are widely applied. They have become an indispensable reference point and a common denominator for the vast majority of Member States that have adopted these standards for use in national regulations to enhance safety in nuclear power generation, research reactors and fuel cycle facilities as well as in nuclear applications in medicine, industry, agriculture and research.

The IAEA safety standards are based on the practical experience of its Member States and produced through international consensus. The involvement of the members of the Safety Standards Committees, the Nuclear Security Guidance Committee and the Commission on Safety Standards is particularly important, and I am grateful to all those who contribute their knowledge and expertise to this endeavour.

The IAEA also uses these safety standards when it assists Member States through its review missions and advisory services. This helps Member States in the application of the standards and enables valuable experience and insight to be shared. Feedback from these missions and services, and lessons identified from events and experience in the use and application of the safety standards, are taken into account during their periodic revision.

I believe the IAEA safety standards and their application make an invaluable contribution to ensuring a high level of safety in the use of nuclear technology. I encourage all Member States to promote and apply these standards, and to work with the IAEA to uphold their quality now and in the future.

THE IAEA SAFETY STANDARDS

BACKGROUND

Radioactivity is a natural phenomenon and natural sources of radiation are features of the environment. Radiation and radioactive substances have many beneficial applications, ranging from power generation to uses in medicine, industry and agriculture. The radiation risks to workers and the public and to the environment that may arise from these applications have to be assessed and, if necessary, controlled.

Activities such as the medical uses of radiation, the operation of nuclear installations, the production, transport and use of radioactive material, and the management of radioactive waste must therefore be subject to standards of safety.

Regulating safety is a national responsibility. However, radiation risks may transcend national borders, and international cooperation serves to promote and enhance safety globally by exchanging experience and by improving capabilities to control hazards, to prevent accidents, to respond to emergencies and to mitigate any harmful consequences.

States have an obligation of diligence and duty of care, and are expected to fulfil their national and international undertakings and obligations.

International safety standards provide support for States in meeting their obligations under general principles of international law, such as those relating to environmental protection. International safety standards also promote and assure confidence in safety and facilitate international commerce and trade.

A global nuclear safety regime is in place and is being continuously improved. IAEA safety standards, which support the implementation of binding international instruments and national safety infrastructures, are a cornerstone of this global regime. The IAEA safety standards constitute a useful tool for contracting parties to assess their performance under these international conventions.

THE IAEA SAFETY STANDARDS

The status of the IAEA safety standards derives from the IAEA's Statute, which authorizes the IAEA to establish or adopt, in consultation and, where appropriate, in collaboration with the competent organs of the United Nations and with the specialized agencies concerned, standards of safety for protection of health and minimization of danger to life and property, and to provide for their application.

With a view to ensuring the protection of people and the environment from harmful effects of ionizing radiation, the IAEA safety standards establish fundamental safety principles, requirements and measures to control the radiation exposure of people and the release of radioactive material to the environment, to restrict the likelihood of events that might lead to a loss of control over a nuclear reactor core, nuclear chain reaction, radioactive source or any other source of radiation, and to mitigate the consequences of such events if they were to occur. The standards apply to facilities and activities that give rise to radiation risks, including nuclear installations, the use of radiation and radioactive sources, the transport of radioactive material and the management of radioactive waste.

Safety measures and security measures[1] have in common the aim of protecting human life and health and the environment. Safety measures and security measures must be designed and implemented in an integrated manner so that security measures do not compromise safety and safety measures do not compromise security.

The IAEA safety standards reflect an international consensus on what constitutes a high level of safety for protecting people and the environment from harmful effects of ionizing radiation. They are issued in the IAEA Safety Standards Series, which has three categories (see Fig. 1).

Safety Fundamentals

Safety Fundamentals present the fundamental safety objective and principles of protection and safety, and provide the basis for the safety requirements.

Safety Requirements

An integrated and consistent set of Safety Requirements establishes the requirements that must be met to ensure the protection of people and the environment, both now and in the future. The requirements are governed by the objective and principles of the Safety Fundamentals. If the requirements are not met, measures must be taken to reach or restore the required level of safety. The format and style of the requirements facilitate their use for the establishment, in a harmonized manner, of a national regulatory framework. Requirements, including numbered 'overarching' requirements, are expressed as 'shall' statements. Many requirements are not addressed to a specific party, the implication being that the appropriate parties are responsible for fulfilling them.

Safety Guides

Safety Guides provide recommendations and guidance on how to comply with the safety requirements, indicating an international consensus that it

---

[1]   See also publications issued in the IAEA Nuclear Security Series.

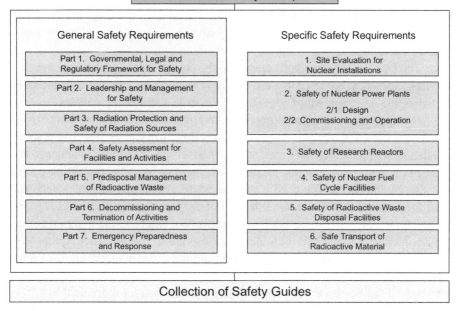

FIG. 1. The long term structure of the IAEA Safety Standards Series.

is necessary to take the measures recommended (or equivalent alternative measures). The Safety Guides present international good practices, and increasingly they reflect best practices, to help users striving to achieve high levels of safety. The recommendations provided in Safety Guides are expressed as 'should' statements.

APPLICATION OF THE IAEA SAFETY STANDARDS

The principal users of safety standards in IAEA Member States are regulatory bodies and other relevant national authorities. The IAEA safety standards are also used by co-sponsoring organizations and by many organizations that design, construct and operate nuclear facilities, as well as organizations involved in the use of radiation and radioactive sources.

The IAEA safety standards are applicable, as relevant, throughout the entire lifetime of all facilities and activities — existing and new — utilized for peaceful purposes and to protective actions to reduce existing radiation risks. They can be used by States as a reference for their national regulations in respect of facilities and activities.

The IAEA's Statute makes the safety standards binding on the IAEA in relation to its own operations and also on States in relation to IAEA assisted operations.

The IAEA safety standards also form the basis for the IAEA's safety review services, and they are used by the IAEA in support of competence building, including the development of educational curricula and training courses.

International conventions contain requirements similar to those in the IAEA safety standards and make them binding on contracting parties. The IAEA safety standards, supplemented by international conventions, industry standards and detailed national requirements, establish a consistent basis for protecting people and the environment. There will also be some special aspects of safety that need to be assessed at the national level. For example, many of the IAEA safety standards, in particular those addressing aspects of safety in planning or design, are intended to apply primarily to new facilities and activities. The requirements established in the IAEA safety standards might not be fully met at some existing facilities that were built to earlier standards. The way in which IAEA safety standards are to be applied to such facilities is a decision for individual States.

The scientific considerations underlying the IAEA safety standards provide an objective basis for decisions concerning safety; however, decision makers must also make informed judgements and must determine how best to balance the benefits of an action or an activity against the associated radiation risks and any other detrimental impacts to which it gives rise.

DEVELOPMENT PROCESS FOR THE IAEA SAFETY STANDARDS

The preparation and review of the safety standards involves the IAEA Secretariat and five Safety Standards Committees, for emergency preparedness and response (EPReSC) (as of 2016), nuclear safety (NUSSC), radiation safety (RASSC), the safety of radioactive waste (WASSC) and the safe transport of radioactive material (TRANSSC), and a Commission on Safety Standards (CSS) which oversees the IAEA safety standards programme (see Fig. 2).

All IAEA Member States may nominate experts for the Safety Standards Committees and may provide comments on draft standards. The membership of the Commission on Safety Standards is appointed by the Director General and includes senior governmental officials having responsibility for establishing national standards.

A management system has been established for the processes of planning, developing, reviewing, revising and establishing the IAEA safety standards.

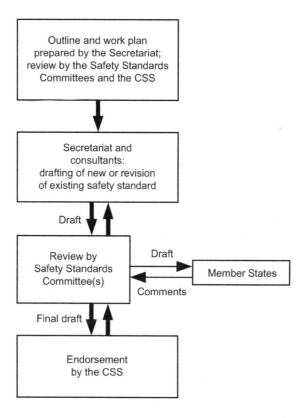

FIG. 2. The process for developing a new safety standard or revising an existing standard.

It articulates the mandate of the IAEA, the vision for the future application of the safety standards, policies and strategies, and corresponding functions and responsibilities.

INTERACTION WITH OTHER INTERNATIONAL ORGANIZATIONS

The findings of the United Nations Scientific Committee on the Effects of Atomic Radiation (UNSCEAR) and the recommendations of international expert bodies, notably the International Commission on Radiological Protection (ICRP), are taken into account in developing the IAEA safety standards. Some safety standards are developed in cooperation with other bodies in the United Nations system or other specialized agencies, including the Food and Agriculture Organization of the United Nations, the United Nations Environment Programme, the International Labour Organization, the OECD Nuclear Energy Agency, the Pan American Health Organization and the World Health Organization.

INTERPRETATION OF THE TEXT

Safety related terms are to be understood as defined in the IAEA Safety Glossary (see https://www.iaea.org/resources/safety-standards/safety-glossary). Otherwise, words are used with the spellings and meanings assigned to them in the latest edition of The Concise Oxford Dictionary. For Safety Guides, the English version of the text is the authoritative version.

The background and context of each standard in the IAEA Safety Standards Series and its objective, scope and structure are explained in Section 1, Introduction, of each publication.

Material for which there is no appropriate place in the body text (e.g. material that is subsidiary to or separate from the body text, is included in support of statements in the body text, or describes methods of calculation, procedures or limits and conditions) may be presented in appendices or annexes.

An appendix, if included, is considered to form an integral part of the safety standard. Material in an appendix has the same status as the body text, and the IAEA assumes authorship of it. Annexes and footnotes to the main text, if included, are used to provide practical examples or additional information or explanation. Annexes and footnotes are not integral parts of the main text. Annex material published by the IAEA is not necessarily issued under its authorship; material under other authorship may be presented in annexes to the safety standards. Extraneous material presented in annexes is excerpted and adapted as necessary to be generally useful.

CONTENTS

1. INTRODUCTION

BACKGROUND

1.1. This Safety Guide provides recommendations on how to meet the requirements established in IAEA Safety Standards Series No. SSR-1, Site Evaluation for Nuclear Installations [1], in relation to the evaluation of hazards generated by earthquakes affecting nuclear power plants and other nuclear installations.

1.2. This Safety Guide supersedes the 2010 version of IAEA Safety Standards Series No. SSG-9, Seismic Hazards in Site Evaluation for Nuclear Installations[1]. This Safety Guide takes into account feedback from Member States on the application of the 2010 version of SSG-9. In particular, the modifications incorporated into this Safety Guide reflect the following:

(a) Progress in practice and research relating to the evaluation of seismic hazards, as well as in the regulatory practice of Member States, considering lessons from recent strong earthquakes that affected nuclear installations;
(b) Recent technical developments and new regulatory requirements relating to risk informed and performance based approaches to assessing the safety of nuclear installations;
(c) Experience and results from seismic hazard assessments performed for the evaluation of new and existing sites for nuclear installations in Member States;
(d) More consistent treatment of seismically induced geological and geotechnical hazards and concomitant events;
(e) A more consistent approach to considering the diversity of professional judgement by experts and the treatment of the uncertainties involved in evaluating seismic hazards.

1.3. This Safety Guide also provides a clearer separation between the process for assessing the seismic hazards at a specific site and the process for defining the related basis for design and evaluation of the nuclear installation. Thus, it bridges gaps and avoids undue overlapping of the two processes, which correspond to and are performed at different stages of the lifetime of the nuclear installation.

---

[1] INTERNATIONAL ATOMIC ENERGY AGENCY, Seismic Hazards in Site Evaluation for Nuclear Installations, IAEA Safety Standards Series No. SSG-9, IAEA, Vienna (2010).

OBJECTIVE

1.4. The objective of this Safety Guide is to provide recommendations on how to meet the requirements established in SSR-1 [1] in relation to the evaluation of hazards generated by earthquakes that might affect a nuclear installation site and, in particular, on how to determine the following:

(a) The vibratory ground motion[2] hazards necessary to establish the design basis ground motions and other relevant parameters for the design and safety assessment of both new and existing nuclear installations;
(b) The potential for, and the rate of, fault displacement phenomena that could affect the feasibility of a site for a new nuclear installation or the safe operation of an existing installation at a site;
(c) The earthquake parameters necessary for assessing the associated geological and geotechnical hazards (e.g. soil liquefaction, landslides, differential settlements, collapse due to cavities and subsidence phenomena) and concomitant events (e.g. external flooding phenomena such as tsunamis and fires).

1.5. This Safety Guide is intended for use by regulatory bodies responsible for establishing regulatory requirements and by operating organizations directly responsible for the evaluation of seismic hazards at a nuclear installation site.

SCOPE

1.6. The recommendations in this Safety Guide are intended to be used for the evaluation of seismic hazards for nuclear installations in any seismotectonic environment.

1.7. This Safety Guide addresses all types of nuclear installation as defined in the IAEA Safety Glossary [2], as follows:

(a) Nuclear power plants;
(b) Research reactors (including subcritical and critical assemblies) and any adjoining radioisotope production facilities;
(c) Storage facilities for spent fuel;
(d) Facilities for the enrichment of uranium;

---

[2] In this Safety Guide, the terms 'vibratory ground motion' and 'ground motion' are synonymous. In some States, vibratory ground motion is called 'earthquake ground motion' or 'seismic ground motion'.

2

(e)   Nuclear fuel fabrication facilities;
(f)   Conversion facilities;
(g)   Facilities for the reprocessing of spent fuel;
(h)   Facilities for the predisposal management of radioactive waste arising from nuclear fuel cycle facilities;
(i)   Nuclear fuel cycle related research and development facilities.

1.8. The recommendations for nuclear power plants are applicable to other nuclear installations by means of a graded approach, whereby these recommendations can be customized to suit the needs of nuclear installations of different types in accordance with the potential radiological consequences of their failure when subjected to seismic loads. The recommended approach is to start with the recommendations for nuclear power plants and to modify the application of those recommendations until they are commensurate with installations with which lesser radiological consequences are associated. If no grading is performed, the recommendations relating to nuclear power plants should be applied to other types of nuclear installation. The level of detail and the effort devoted to evaluating the seismic hazards at existing installation sites should be commensurate with a number of additional factors (e.g. the time remaining until the installation is expected to be shut down, the stage of site remediation, the severity of the seismic hazards where the site is located).

1.9. For the purpose of this Safety Guide, existing nuclear installations are installations that are (a) at the operational stage (including long term operation and extended temporary shutdown periods); (b) at a pre-operational stage for which the construction of structures, the manufacturing, installation and/or assembly of components and systems, and commissioning activities are significantly advanced or fully completed; or (c) at a temporary shutdown, permanent shutdown or decommissioning stage, with radioactive material still within the installation (e.g. in the reactor core or the spent fuel pool).

1.10. Earthquakes generate several direct and indirect phenomena, from vibratory ground motions to associated geological and geotechnical hazards, such as permanent ground displacement (e.g. soil liquefaction, slope instability, tectonic and non-tectonic subsidence, cavities leading to ground collapse, differential settlements), to subsequent concomitant events such as seismically induced fires and floods. This Safety Guide provides guidance on how to consistently characterize and define the seismic parameters necessary for evaluating the geological and geotechnical hazards and concomitant events as described in IAEA Safety Standards Series No. NS-G-3.6, Geotechnical Aspects of Site Evaluation and Foundations for Nuclear Power Plants [3], and IAEA Safety Standards Series

No. SSG-18, Meteorological and Hydrological Hazards in Site Evaluation for Nuclear Installations [4], respectively.

1.11. This Safety Guide addresses aspects relating to the evaluation of hazards generated by earthquakes that might affect the site. This evaluation will be performed at the site selection and/or site evaluation stages, possibly prior to the availability of information relating to the design characteristics of the nuclear installation, or during the operation stage of an existing nuclear installation. Thus, the seismic hazards may need to be determined independently of the characteristics of the nuclear installation that is to be installed. Recommendations for the determination of the appropriate basis for the design and evaluation of a nuclear installation through the use and application of appropriate criteria are provided in IAEA Safety Standards Series No. SSG-67, Seismic Design for Nuclear Installations [5].

STRUCTURE

1.12. Recommendations of a general nature are provided in Section 2. Section 3 provides recommendations on the acquisition of a database containing the information needed to evaluate and address all hazards associated with earthquakes. Section 4 covers the use of this database for the development of seismic source models specific to the site of the nuclear installation. Section 5 provides recommendations on available methods for conducting vibratory ground motion analysis. Section 6 provides recommendations on probabilistic and deterministic methods for evaluating vibratory ground motion hazards. Section 7 presents methods for evaluating the potential for fault displacement. Section 8 provides recommendations relating to parameters from the vibratory ground motion analysis, fault displacement and other associated seismic hazards.

1.13. Sections 3–8 focus primarily on nuclear power plants. Section 9 provides recommendations on the application of a graded approach in evaluating seismic hazards for nuclear installations other than nuclear power plants. Section 10 addresses the application of the management system, including project management and peer reviews. The Annex provides an example of typical output deriving from probabilistic seismic hazard analyses.

2. GENERAL ASPECTS OF SEISMIC HAZARD ASSESSMENT

2.1. The following requirements are established in SSR-1 [1]:

"**Requirement 1: Safety objective in site evaluation for nuclear installations**

"**The safety objective in site evaluation for nuclear installations shall be to characterize the natural and human induced external hazards that might affect the safety of the nuclear installation, in order to provide adequate input for demonstration of protection of people and the environment from harmful effects of ionizing radiation.**"

．．．．．．．

"**Requirement 15: Evaluation of fault capability**

"**Geological faults larger than a certain size and within a certain distance of the site and that are significant to safety shall be evaluated to identify whether these faults are to be considered capable faults. For capable faults, potential challenges to the safety of the nuclear installation in terms of ground motion and/or fault displacement hazards shall be evaluated.**"

．．．．．．．

"**Requirement 16: Evaluation of ground motion hazards**

"**An evaluation of ground motion hazards shall be conducted to provide the input needed for the seismic design or safety upgrading of the structures, systems and components of the nuclear installation, as well as the input for performing the deterministic and/or probabilistic safety analyses necessary during the lifetime of the nuclear installation.**"

In accordance with these requirements and in line with recognized international practice, the geological, geophysical and seismological characteristics of the geographical region around the site and the geotechnical characteristics of the site area should be investigated to evaluate the seismic hazards at the nuclear installation site.

2.2. The size of the region to be analysed should be determined on the basis of the types, magnitudes and distances from the source to the site of potentially hazardous phenomena generated by earthquakes that might have an impact on the safety of the nuclear installation. Thus, the region should be of sufficient extent to include all seismic sources that could reasonably be expected to contribute to the seismic hazards at the site. The region will not necessarily have predetermined uniform dimensions, and it should be defined on the basis of the specific conditions associated with the site and the region. If necessary, the region should include areas extending beyond national borders as well as relevant offshore areas.

2.3. The size of the region to be investigated, the type of information and data to be collected, and the scope and detail of the investigations to be performed should be defined at the beginning of the seismic hazard assessment project. The acquired database should be sufficient for characterizing, from a seismotectonic point of view, features relevant to the seismic hazard assessment that are located in other States or in offshore areas.

2.4. The evaluation of seismic hazards for a nuclear installation site should be done through the implementation of a specific project plan for which clear and detailed objectives are defined, and with a project organization and structure that provides for coherency and consistency in the database and a reasonable basis on which to compare results for all types of seismic hazard. This project plan should include an independent peer review. It should be carried out by a multidisciplinary team of experts, including geologists, seismologists, geophysicists, seismic hazard specialists, engineers and possibly other experts (e.g. historians) as necessary. The members of the team for the seismic hazard assessment project and the independent peer review should demonstrate expertise and experience commensurate with their role in the project. Figure 1 shows the seismic hazard assessment process as a whole and the general steps and sequence to be followed.

2.5. The general approach to seismic hazard assessment should be directed towards the realistic identification, quantification, treatment and reduction of uncertainties through all stages of the project. Experience shows that the most effective way of achieving this is to collect sufficient reliable and relevant site specific data. There is generally a compromise between the time and effort needed to compile a detailed, reliable and relevant database and the degree of uncertainty that should be taken into consideration at each step of the process. Thus, applying a lower level of effort in developing the database for characterization of the seismic sources, fault capabilities and ground motions will result in increased uncertainty in the final results obtained.

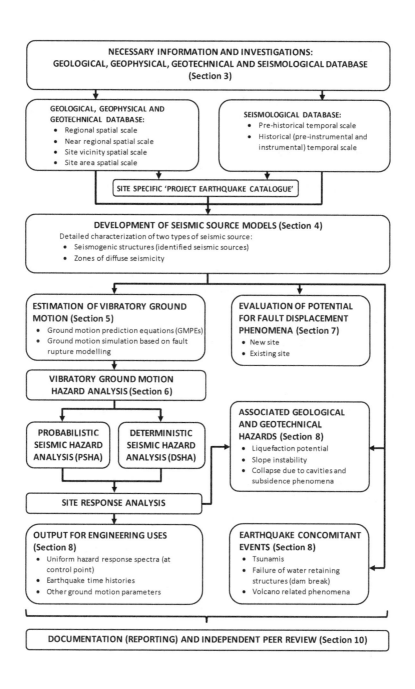

FIG. 1. Flow chart showing the seismic hazard assessment process for nuclear installations.

2.6. Therefore, an adequate method for identification, quantification and treatment of the uncertainties should be formulated at the beginning of the project. In general, significant uncertainties are associated with the seismic hazard assessment process. Basically, two types of uncertainty are identified for practical application in seismic hazard assessment: (i) the aleatory variability of the seismic process, which is inherent in phenomena that occur in a random manner and as such cannot be reduced, even by collecting more data, and (ii) the epistemic uncertainty, which is attributable to incomplete knowledge about a phenomenon (therefore affecting the ability to model it) and which can be reduced through the acquisition of additional data (including site specific data), further research and interaction between experts considering the diversity of their professional judgement [2].[3]

2.7. Site specific, sufficient and reliable data should be collected in the seismic hazard assessment process. However, part of the data used indirectly in the seismic hazard analysis might not be site specific (in particular, the data on strong motions used to develop ground motion prediction equations (GMPEs)). Therefore, relevant uncertainties should be taken into consideration.

2.8. One of the main sources of epistemic uncertainty in seismic hazard assessment is the differences in interpretation of the available data owing to the diversity of professional judgement of the experts participating in the hazard assessment process. Care should be taken to avoid bias in these interpretations. Expert judgement should not be used as a substitute for acquiring new data. The project team for the seismic hazard assessment should evaluate, without bias, all hypotheses and models supported by the data compiled and should then develop an integrated model that takes into account both existing knowledge and uncertainties in the data. Where it is required to evaluate much longer periods (lower exceedance frequencies) than the data permit, knowledge of the regional and local geodynamics and neotectonics can support the use of expert judgement.

2.9. Structured expert interactions should be employed to avoid artificial influence of uncertainty estimates on the results. To address the diversity of scientific interpretations, the centre, body and range of the technically defensible interpretations should be properly captured [6]. For this purpose, multidisciplinary teams of experts with appropriate qualifications in each of the relevant areas should be involved in developing a model that robustly represents the epistemic

---

[3] Seismic hazard analyses assume that the geological processes are stationary because the timescale over which the analysis is needed for a site (a few decades) is much shorter than the timescale over which geodynamic changes take place.

uncertainties relating to methods and models employed in the seismic hazard assessment. Where an approach makes use of expert elicitation, care should be exercised to ensure that professional judgements made by experts are supported, so far as is practicable, by the available earth science data. Also, adequate consideration should be given to uncertainties using suitable (e.g. conservative, best estimate) and credible models, methods and scenarios — based on the concept of technically defensible interpretations — as appropriate for the evaluation framework (i.e. deterministic or probabilistic) and the target confidence levels. The composition of the peer review panel should also reflect the size and complexity of the project generally.

2.10. A set of quality assurance documents should be prepared and properly updated during the seismic hazard assessment process. All technical references used in the process will be useful, since the guidance they provide might be interpreted in different ways. An unambiguous set of project specific quality documents (e.g. quality plan, work plan and procedures) should be prepared so that the set contains all the criteria applicable to the project at hand; documentation recording all expert interpretations should also be included. More detailed recommendations on this topic are provided in Section 10.

2.11. As indicated in para. 2.8, uncertainties that cannot be reduced by means of site specific investigations (e.g. uncertainties arising from the use of GMPEs derived for other parts of the world) do not permit hazard values to decrease below certain threshold values. For this reason, and irrespective of any lower apparent seismic hazard associated with the site, a minimum vibratory ground motion level should be recognized as the lower limit to be used for seismic design, safety assessment and/or seismic safety evaluation of any nuclear installation, and that minimum level should be adopted when applying the recommendations in SSG-67 [5].

3. DATABASE OF INFORMATION AND INVESTIGATIONS

GENERAL

3.1. A comprehensive and integrated database of geological, geophysical, geotechnical and seismological information should be compiled in a coherent form for use in evaluating and resolving issues relating to hazards generated by earthquakes.

3.2. It should be ensured that each element of each individual database has been investigated as fully as possible before integration of the various elements into a unique consolidated database is attempted. The integrated database should include all relevant information, not only geological, geophysical, geotechnical and seismological data but also any other information relevant to evaluating the vibratory ground motion, the fault displacement phenomena, the associated geological and geotechnical hazards, and the concomitant events affecting the site.

3.3. The data and information to be acquired for the geological, geophysical, geotechnical and seismological database should cover a geographical region and a temporal scale commensurate with the potential of the seismic hazards to affect the safety of the nuclear installation at the site.

3.4. In relation to the geographical area of interest to be investigated, SSR-1 [1] states:

"**Requirement 5: Site and regional characteristics**

"**The site and the region shall be investigated with regard to the characteristics that could affect the safety of the nuclear installation and the potential radiological impact of the nuclear installation on people and the environment.**

"4.12. Natural phenomena as well as human activities in the region with the potential to induce hazards at the site that might affect the safety of the nuclear installation shall be identified and evaluated. The extent of this evaluation shall be commensurate with the safety significance of the potential hazards at the site.

"4.13. The characteristics of the natural environment in the region that could be affected by the potential radiological impact of the nuclear installation shall be investigated and assessed, for all operational states and accident conditions and for all stages of the lifetime of the nuclear installation (see Section 6 [of SSR-1 [1]]).

"4.14. The size of the region to be investigated shall be defined for each of the natural and human induced external hazards. Both the magnitude of the hazard and the distance from the source of the hazard to the site shall be considered in determining the size of the region to be investigated. For certain natural external events, such as tsunamis and volcanic phenomena, it shall be ensured that the size of the region that is investigated is sufficiently large to address the potential effects at the site.

"4.15.   The site and the region shall be studied to evaluate the present and foreseeable future characteristics that could have an impact on the safety of the nuclear installation. This includes potential changes in the severity and/or the frequency of natural external events, as well as changes in the population distribution in the region, the present and future use of land and water, the further development of existing nuclear installations or the construction of other facilities that could affect the safety of the nuclear installation or the feasibility of planning effective emergency response actions."

3.5.   In relation to the temporal scale of the investigations, SSR-1 [1] states:

"Requirement   14:   Data   collection   in   site   evaluation   for nuclear installations

"The data necessary to perform an assessment of natural and human induced external hazards and to assess both the impact of the environment on the safety of the nuclear installation and the impact of the nuclear installation on people and the environment shall be collected.

....

"4.47.   Information and records, if available, of the occurrence and severity of important prehistoric, historical and recent natural phenomena shall be obtained as appropriate for the hazard to be evaluated and shall be analysed for reliability, accuracy, temporal and spatial relevance, and completeness."

3.6.   The size of the geographical area at the regional scale for which the geological, geophysical, geotechnical and seismological database should be compiled may differ depending on the geological and tectonic setting, and the recommendations provided in para. 2.3 should be used to define the appropriate size of the region to be investigated.

3.7.   The geological, geophysical and geotechnical investigations for evaluating the seismic hazards at the site should be conducted on four spatial geographical scales — regional, near regional, site vicinity and site area — leading to progressively more detailed investigations, data and information. The detail and type of these data are determined by the different spatial geographical scales. The first three scales of investigation lead, primarily, to progressively more detailed geological and geophysical data and information. The site area investigations are mainly aimed at developing the geophysical and geotechnical database for evaluation of vibratory ground motion and fault displacement.

3.8. With the completion of the geological, geophysical and geotechnical investigations at the four spatial scales, all seismogenic features that have been identified and characterized, including assessment of the uncertainties for all fault parameters, should be documented finally and in a systematic way to ensure consistency and completeness, so that similar attributes for all seismic sources can be compiled in the 'project fault catalogue' (also known as the 'project fault portfolio').

3.9. The seismological database should include all available information and data on earthquake events that have occurred in the region, and such information and data should cover the pre-historical and historical temporal scales. The historical temporal scale should be further subdivided into pre-instrumental and instrumental periods.

3.10. In offshore regions and other areas for which seismological data are poor, adequate investigations should be conducted to fully analyse the tectonic characteristics of the region and to compensate for any lack of or deficiency in the seismological data.

3.11. In investigations to evaluate the potential for earthquake generated tsunamis, the geological and seismological investigations should also include the study of seismic sources located at very great distances from the site. Thus, the sources of earthquakes that can generate relevant seismic hazards and relevant tsunami hazards at the site might not be the same. For tsunamis generated by earthquake induced submarine landslides, the models used to calculate the ground motion inducing the landslide should be consistent with those models used in the seismic hazard assessment for the nuclear installation.

3.12. New techniques that have recently emerged in the acquisition and processing of data (e.g. remote sensing, age dating, use of dense seismic observation networks) for the identification and characterization of seismic sources should be implemented. It is also possible that new types of data might be generated as a result of these technological developments. While it is recommended that state of the art, new, updated and recognized technological developments be implemented, such developments should first be checked for adequacy and effectiveness before being used in a nuclear installation site evaluation project.

3.13. As earthquakes produce observable effects on the environment, palaeoseismological studies should be performed, as necessary, at any of the four spatial scales to achieve the following:

(a) To identify the seismogenic structures on the basis of recognition of effects of past earthquakes in the region.
(b) To improve the completeness of earthquake catalogues for large events, using identification and age dating of geological markers such as fossils. For example, observations of trenching across the identified potential capable faults may be useful in estimating the amount of displacement (e.g. from the thickness of colluvial wedges) and its rate of occurrence (e.g. by age dating of the sediments). Also, studies of palaeo-liquefaction, palaeo-landslides and palaeo-tsunamis can provide evidence of the recurrence and intensity of earthquakes.
(c) To estimate the potential maximum magnitude (and the associated uncertainty) of a given seismogenic structure, typically based on the maximal dimensions of the structure and the displacement per event (estimated from the trenching) as well as the cumulative effect of all seismogenic structures (estimated from the seismic landscape[4]).

3.14. To achieve consistency in the presentation of information, the data should be compiled in a geographical information system with adequate metadata. All data should be stored in a uniform reference frame to facilitate comparison and integration.

3.15. When a seismic hazard assessment is performed during the lifetime of the nuclear installation (e.g. for a periodic safety review or a seismic probabilistic safety assessment), the existing database should be updated in accordance with the recommendations provided in paras 3.1–3.14 above as part of the seismic hazard re-evaluation process.

---

[4] The seismic landscape is the cumulative geomorphic and stratigraphic effect of the signs left on an area's physical environment by its past earthquakes over a geologically recent time interval.

GEOLOGICAL, GEOPHYSICAL AND GEOTECHNICAL DATABASE

Regional investigations

3.16. The purpose of obtaining geological and geophysical data on a regional scale is to provide knowledge of the general geodynamic setting of the region and the current tectonic regime, as well as to identify and characterize those geological features evaluated from investigations, such as lithology, geomorphology, stratigraphy and fault investigations, that might influence or relate to the seismic hazard at the site.

3.17. Thus, the extent of the geographical area of interest at a regional scale should be defined in accordance with the recommendations provided in para. 3.6 and by considering the potential sources of all hazards generated by earthquakes that might affect the safety of the nuclear installations at the selected site. The size of the region to be investigated when assessing vibratory ground motion hazards should be large enough to incorporate all seismogenic structures that could affect the nuclear installation: the extent of this region is typically a few hundred kilometres in radius, or in keeping with the national requirements of the State.

3.18. Existing data from any type of published or unpublished geological or geophysical source (e.g. data from the literature; data on the country as a whole; remote sensing data; data derived from existing galleries or road cuts, geophysical surveys or geotechnical characteristics) should be searched and, if necessary, confirmed by direct observation through geological field reconnaissance visits.

3.19. Where existing data are insufficient to properly characterize the identified potential geological features relevant to the seismic hazard at the site, further investigations should be considered; if necessary, these data should be interpreted using reasonable and defensible hypotheses. It may be necessary to complement the data by acquiring new geological and geophysical data of sufficient detail, similar to the level of detail for the near region. If needed, identification and analysis of geological and geomorphological evidence (i.e. palaeoseismology; see para. 3.13) of pre-historical and historical earthquakes, including geodynamic investigations, should also be performed for this purpose.

3.20. The data collected at the regional scale should have a resolution that can reveal any features considered to be significant for the analysis of the seismic hazard, with appropriate cross-sections. The collected data and the results obtained should have a resolution consistent with maps at the appropriate scale. The data should be organized in the project geographical information system within the layer of regional scale information, and a summary report should be prepared to describe the studies and investigations performed and results obtained, particularly in relation to the seismogenic structures identified at this stage of the studies.

Near regional investigations

3.21. Geological, geophysical and geotechnical investigations should be conducted in more detail in the near region to provide more specific information than that available from the regional studies, with the following objectives:

(a) To define the seismotectonic characteristics of the near region;
(b) To determine the most recent movements of the seismogenic structures and/or potential capable faults identified in the near region;
(c) To determine the amount and nature of displacements, rates of activity and evidence relating to the segmentation of such seismogenic structures.

3.22. The near regional studies should include a geographical area typically not less than 25 km in radius from the site boundary, although this dimension should be adjusted to reflect local seismotectonic conditions. For new nuclear installation sites for which the exact layout of the buildings and structures has not been defined, the near regional area should be defined from the boundary of the prospective site area.

3.23. These more detailed geological, geophysical and geotechnical investigations should supplement the published and unpublished information already collected for the near regional area, and they should include a definition of the stratigraphy, structural geology and tectonic history of the near region. The tectonic history should be thoroughly defined for the current tectonic regime, the length of which will depend on the rate of tectonic activity. For example, for studies to assess fault capability, the tectonic information through the Upper Pleistocene to the Holocene (i.e. the present) may be adequate for high seismic regions, while for low seismic regions information through the Pliocene to the Holocene may be necessary.

3.24. In general, for the near regional scale as a whole, the following investigations should be performed in accordance with the procedures and methods established by recognized applicable industry codes and standards. Some of these investigations should be performed specifically around the identified geological features that have the potential to generate seismic hazards at the site:

(a) Geomorphological studies of Quaternary formations or land forms, such as terrace analysis and pedological and sedimentological studies, using well recognized remote sensing image techniques (e.g. aerial and satellite photographs and/or images, light detection and ranging (lidar)). Bathymetric information should also be obtained for geomorphological investigation in dealing with offshore areas for sites located on or near a coastline.

(b) Field geological mapping to identify geomorphology at the scale necessary for the near regional studies.

(c) Subsurface data derived from borehole and geophysical investigations — such as high resolution seismic reflection and/or refraction profiles, and gravimetric, electric and magnetic tomography techniques — to spatially characterize the identified seismogenic structures considered relevant in terms of their geometry, extent and rate of deformation. The use of heat flow data may also be necessary.

(d) Geochronological dating, using recognized, reliable and applicable techniques with appropriate care for stratigraphic purposes.

(e) Data derived from geodetic methods — such as global navigation satellite systems (GNSSs), including the Global Positioning System (GPS), and interferometry images — and strain rate measurements to assess the ongoing rate and type of tectonic deformation.

(f) Hydrogeological investigations using new and existing boreholes, wells and other techniques to define the geometry, physical and chemical properties, and steady state behaviour (e.g. water table depth, recharge rate, transmissivity) of all aquifers at the scale necessary for the near regional studies.

(g) Palaeoseismological and trenching investigations, as needed, based on the analysis of the data and results obtained from the studies listed in (a)–(f) above.

(h) Collection of instrumental data from seismic monitoring networks; see paras 3.54–3.59.

3.25. Investigations should be made in sufficient detail that the causes of each relevant geological and geomorphological feature (e.g. topographical or structural features found in aerial photographs, remote sensing imagery or geophysical data) can be properly included in a reasonable model postulated for the recent geological evolution of the area.

3.26. The data collected and the results obtained from the investigations performed at the near regional scale should have a resolution consistent with maps at a scale of typically 1:50 000, or larger, with appropriate cross-sections. Digital elevation models should also be part of the results obtained from this task. The data should be organized in the project geographical information system within the layer of near regional scale information. A summary report should be prepared to describe the studies and investigations performed, the evaluation of information for inclusion in the models, and the results obtained, particularly in relation to the seismogenic structures further identified and characterized at this stage of the studies.

Site vicinity investigations

3.27. In addition to the information collected at the regional and near regional scales, more specific geological, geophysical and geotechnical studies should be conducted in the site vicinity with the objective of providing a more completed database for this smaller area regarding the definition and characterization in greater detail of the neotectonic history of the identified seismogenic structures (e.g. faults), especially to determine the potential for and the rate of fault displacement at the site (fault capability) and to identify conditions of potential geological and/or geotechnical instability and associated earthquake generated hazards that might affect the nuclear installation.

3.28. Site vicinity studies should cover a geographical area sufficient to encompass all faults and other seismotectonic features requiring detailed geophysical investigation; this area is typically not less than 5 km (see para. 1.12 of SSR-1 [1]) in radius from the site boundary. For new nuclear installation sites for which the exact layout of the buildings and structures has not been defined, the 5 km radius should be defined from the boundary of the prospective site area.

3.29. Geological, geophysical and geotechnical investigations of the site vicinity should be planned and performed in greater detail than those performed for the near regional scale and should be consistent with the tectonic environment and geological features identified and characterized in previous scale studies (i.e. at the regional and near regional scales). To this end, more detailed geophysical and geotechnical investigations should be undertaken in the site vicinity, including the drilling of boreholes of an adequate number and depth, as well as sampling and laboratory testing.

3.30. These detailed investigations should be performed in accordance with the procedures and methods established by recognized applicable industry codes and standards, and as a result the following data should be obtained:

(a) Geological map at the site vicinity scale with cross-sections;
(b) Age, type, amount and rate of displacement of all the seismogenic structures identified in the site vicinity;
(c) Identification and characterization of locations potentially exhibiting hazards induced by earthquake (e.g. landslide, subsidence, collapse of subsurface cavities or karstic features, failure of dams or water retaining structures).

3.31. The data collected and the results obtained at the site vicinity scale should have a resolution consistent with maps at a scale of typically 1:5000, or larger, with appropriate cross-sections. Digital elevation models should also be part of the results obtained from this task. The data should be organized in the geographical information system within the layer of site vicinity scale information, and a summary report should be prepared to describe the studies and investigations performed, the evaluation of information for inclusion in the models, and the results obtained, particularly in relation to the seismogenic structures further identified and characterized at this stage of the studies.

Site area investigations

3.32. Additional geological, geophysical, geotechnical and seismological site specific studies should be conducted in the nuclear installation site area with the primary objective of providing (a) detailed knowledge for assessing the potential for permanent ground displacement phenomena associated with earthquakes (e.g. surface fault rupture, liquefaction, subsidence or collapse due to subsurface cavities) and (b) information on the static and dynamic properties of rock and soil materials beneath the structure's foundations (e.g. P wave and S wave velocities, seismic quality factor Q,[5] density) to be used in the site response analysis to assess the vibratory ground motions that might affect the safety of the structures, systems and components of the nuclear installation.

3.33. The site area studies should include the entire area covered by the nuclear installation. For a proposed new site for a nuclear installation, at the site evaluation stage the exact layout of the units and/or installations might not yet be known and, for this reason, the entire prospective site area should be considered. For the existing site of an operating nuclear installation for which seismic safety re-evaluation is required, the site area will generally be well defined. If construction is planned for additional nuclear installation units on the existing site area, this should be taken into consideration in defining the extent of the site area.

3.34. Detailed geological, geophysical and geotechnical investigations and studies of the site area should be performed in accordance with the procedures and methods

---

[5] The seismic quality factor Q is a dimensionless factor that quantifies the effects of absorption (anelastic attenuation) of a seismic wave caused by fluid movement and grain boundary friction. Q can be measured experimentally by various techniques and is often characteristic of a particular rock type. Q is inversely proportional to the attenuation coefficient.

established by recognized applicable industry codes and standards and through the use of field and laboratory techniques, as follows:

(a) Geological, geophysical and geotechnical investigations to define the detailed stratigraphy and structure of the area should be conducted. Where practicable, boreholes should be drilled down to the bedrock, and sampling and/or test excavations (including in situ testing), geophysical techniques and laboratory tests should be performed to determine the thickness, depth, dip, and physical and mechanical (static and dynamic) properties of the different subsurface layers as needed by engineering models (e.g. Poisson's ratio, Young's modulus, shear modulus reduction or non-linear properties, dynamic damping properties, density, relative density, shear strength and consolidation characteristics, grain size distribution, P wave and S wave velocities). If necessary, for example in limestone areas, boreholes should also be drilled deep enough to confirm that no cavities or karstic features are underlying the foundations of a nuclear installation.

(b) The data collected in the investigations described in (a) should be enough to indicate whether strata beneath the site are significantly non-horizontal. For example, the soil profile may change across a nuclear installation site as a result of sloping geological layering. In such cases, the subsurface structures across the site may be better modelled as three dimensional, rather than two dimensional, and it may be necessary to enhance the investigations undertaken (e.g. drilling more boreholes) to facilitate the adequate characterization of such sloping geology.

(c) Hydrogeological investigations using boreholes and other techniques should be conducted to define the geometric, physical and chemical properties and steady state behaviour (e.g. water table depth, recharge rate, transmissivity) of all aquifers in the site area, with the specific purpose of determining the stability of soils and how they interact with the foundations of the nuclear installation structures and components.

(d) All the data necessary for assessing the specific site response and the dynamic soil–structure interaction analysis should be acquired in these investigations at the site area. For completeness and efficiency, the investigations described should be integrated with the investigations needed for the dynamic soil–structure interaction, as described in NS-G-3.6 [3] and SSG-67 [5].

3.35. The data collected at the site area scale are typically presented on maps at a scale of 1:500, or larger, with appropriate cross-sections. The data should be organized in the geographical information system within the layer of site area scale information, and a summary report should be prepared to describe the studies and investigations performed, the evaluation of that information for inclusion in the models, and the results obtained, particularly in relation to the seismogenic structures and associated seismic hazards further identified and characterized at this stage of the studies.

SEISMOLOGICAL DATABASE

3.36. To enable reliable characterization of events that occur with very long recurrence periods (or very low annual frequencies of exceedance), the seismological database should include information on past events that might have generated seismic hazards at the site. The database should recognize two types of data relating to two temporal scales — historical and pre-historical — as defined below:

(a) Historical period: the period for which there are documented records of earthquake events. This period is further subdivided as follows:
    (i) Pre-instrumental (or non-instrumental) period: the period before the development and use of instruments to record earthquake parameters;
    (ii) Instrumental period: the period after the development and use of instruments to record earthquake parameters.
(b) Pre-historical period: the period for which there are no documented records of earthquake events. It includes the period in which earthquake evidence might only be retrieved from archaeological sites as described in carvings, paintings, monuments, drawings and other artefacts, including palaeoseismological and geological evidence.

3.37. A specific project earthquake catalogue should be developed from the seismological investigations as an end product of the seismological database. It should include all earthquake related information developed for the project and cover all the temporal scales defined in para. 3.36.

Pre-historical and pre-instrumental historical earthquake data

3.38. All pre-historical and pre-instrumental data on earthquakes should be collected, extending as far back in time as possible. Palaeoseismic and archaeo-seismological information on historical and prehistoric earthquakes should also be collected for such purposes.

3.39. To the extent possible, for each earthquake within these temporal scales, the database should include information on the following:

(a) The date, time and duration of the event;

(b)   The location of the macroseismic epicentre of the event;

(c)   The estimated focal depth of the event;

(d)   The estimated magnitude of the event, including the type of magnitude (e.g. moment magnitude, surface wave magnitude, body wave magnitude, local magnitude, duration magnitude), documentation of the methods used to estimate the magnitude from the macroseismic intensity, and the estimated uncertainty in the magnitude estimate;

(e)   The maximum intensity and, if different, the intensity at the macroseismic epicentre, with a description of local conditions and observed damage;

(f)   The isoseismal contours of the event;

(g)   The intensity of the earthquake at the nuclear installation site, together with any available details of effects on the soil and the landscape;

(h)   Estimates of uncertainty for all the parameters mentioned in (a)–(g) above;

(i)   An assessment of the quality and quantity of data on the basis of which such parameters have been estimated;

(j)   Felt foreshocks and aftershocks;

(k)   The causative fault.

3.40. The intensity scale used in the project earthquake catalogue should be specified, because intensity levels can differ depending on the scale used. The estimates of magnitude and depth for each earthquake should be based on relevant empirical relationships between instrumental data and macroseismic information, which may be developed from the database either directly from observed seismic intensities or by using isoseismals.

Instrumental historical earthquake data

3.41. All available instrumental earthquake data should be collected. Existing information on crustal models should be obtained to locate the epicentres of earthquakes.

3.42. Where sufficient instrumental data exist, for each earthquake the database should include information on the following:

(a)   The date, duration and time of origin of the event;

(b)   The coordinates of the epicentre;

(c)   The focal depth of the event;

(d)   All magnitude determinations, including those on different scales;

(e)   Observed or recorded foreshocks and aftershocks;

(f)   Other information that may be helpful in understanding the seismotectonic regime, such as focal mechanism, seismic moment, stress drop and other source parameters;

(g)   Macroseismic details;

(h)   Fault rupture inhomogeneity such as asperity (or the strong motion generation area), location and size (see Ref. [7] for further details);

(i)   Estimates of uncertainty for each of the parameters mentioned;

(j)   Information on the causative fault, including geometrical features (i.e. length, width, depth, coordinates, strike, dip and rake angles), directivity and duration of rupture;

(k)   Records from both broadband seismometers and strong motion accelerographs with observation station detail.

3.43. Wherever possible, available recordings of regional and local strong ground motion should be collected and used in deriving appropriate ground motion characteristics, as discussed in Section 6.

Project earthquake catalogue

3.44. For a proposed new site of a nuclear installation, a specific project earthquake catalogue should be developed for the entire regional area through four major stages: (1) catalogue compilation; (2) assessment of a uniform size measure to apply to each earthquake (this will include magnitude scale conversions to express all catalogue entries on a single magnitude scale, normally moment magnitude (M_W)); (3) identification of dependent earthquakes (catalogue declustering); and (4) assessment of the completeness of the catalogue as a function of location, time and source size. For sites with existing nuclear installations for which earthquake catalogues are already available, these catalogues should be updated to reflect the newly collected data and information as well as newly available methods.

3.45. When the site specific catalogue of raw pre-historical and historical (including pre-instrumental and instrumental) earthquake data has been compiled, an assessment of the completeness and reliability of the information it contains — particularly in terms of macroseismic intensity, magnitude, date, location and focal depth — should be conducted to verify the records of occurrence of all known earthquakes in the magnitude range considered important in characterizing future seismic hazards. In general, the database will be incomplete for small magnitude events owing to the threshold of recording sensitivity, and it will also be incomplete for large magnitude events owing to their long recurrence intervals (and the comparatively short period of coverage of the catalogues). Appropriate methods should be used to take account of this incompleteness. In general, different periods of completeness should be identified using statistical methods and considering historical and social context.

3.46. When existing catalogues are incorporated and data are transferred from these catalogues to the site specific project earthquake catalogue, the priorities for including one data point rather than another should be established with care. Where data from different existing catalogues are inconsistent or incompatible, clear criteria should be established to govern how such issues are resolved, so that a defensible rationale exists for accepting or rejecting such data.

3.47. If the seismic hazard analysis necessitates that the database is to be composed of independent events (i.e. Poissonian), then a declustering analysis should be performed to identify and separate foreshocks and aftershocks.

3.48. The uncertainties relating to the parameters indicated in the data from pre-historical and historical periods should be identified and quantified to the extent possible. These uncertainties should also be included in the catalogue.

3.49. In summary, prior to the use of the project earthquake catalogue, either to estimate the magnitude–frequency relationship for a seismic source or to estimate the potential maximum magnitude value for each seismic source, thorough evaluation and processing of data in the catalogue should be performed. This evaluation and processing should include the following:

(a) Selection of a consistent magnitude scale for use in the seismic hazard analysis;
(b) Determination of the uniform magnitude of each event in the catalogue on the selected magnitude scale;
(c) Identification of main shocks (i.e. declustering of foreshocks and aftershocks);
(d) Estimation of completeness of the catalogue as a function of magnitude, regional location and time period;
(e) Quality assessment of the derived data, with uncertainty estimates of all parameters.

3.50. All aspects of the development of the earthquake catalogue should be reported to justify the judgements that have been made in compiling it. Specific attention should be paid to the selection of empirical magnitude conversion relations and to the selection of the magnitude scale for all catalogue entries. A comparison of the project catalogue with other similar catalogues relevant to the region should be performed.

3.51. The magnitude scale selected for the catalogue should be consistent with the magnitude scale utilized in the GMPEs used in the vibratory ground motion hazard calculations. In deriving magnitude–frequency relationships, the selected magnitude scale should vary almost linearly with M_W across the magnitude range of interest, to avoid magnitude saturation effects. This approach is consistent with the use of M_W becoming a worldwide standard, owing to its increased use in seismology and the development of GMPEs.

3.52. A magnitude–frequency relationship should be developed for each seismic source. Each magnitude–frequency relationship should include the potential maximum magnitude for which the magnitude–frequency relationship applies.

3.53. Uncertainty in the parameters of the magnitude–frequency relationship should be defined by probability distributions that take into account any correlation between the parameters.

Site specific instrumental data

3.54. To acquire more detailed information on potential seismic sources, it is advantageous to install or have access to a seismic monitoring network system of high sensitivity seismometers. This system should be installed and operated in the near region around the nuclear installation site and within the site itself. The seismometers should have the capability of recording micro-earthquakes and sufficiently high frequencies. The design of the seismic monitoring network system should be suitable for the geological setting and for assessing the seismic hazards at the site. The data obtained from the operation of this system should also be used as a supporting tool in decisions regarding the capability of faults (see Section 7).

3.55. The seismic monitoring network system should be installed for new sites from the very beginning of the site evaluation stage. For existing sites for which such systems were not originally deployed, the seismic monitoring network system should be installed from the beginning of the seismic safety re-evaluation programme. These systems should be operated during the whole lifetime of the nuclear installation.

3.56. The operation and data processing of these seismic monitoring network systems should be linked to any existing regional and/or national seismic monitoring network systems.

3.57. If the selected instrumentation for the seismic monitoring network system cannot adequately record strong motions, several strong motion accelerometers should be collocated with the high sensitivity seismometers to acquire more detailed information on path effects, empirical Green's functions, GMPEs and site responses. In addition, measurement of ambient noise (i.e. micro-tremors) should be deployed, if necessary, to evaluate the site response.

3.58. Earthquakes recorded within and near the seismic monitoring network system should be carefully analysed in connection with seismotectonic studies of the near region.

3.59. The instrumentation used should be appropriately and periodically upgraded and calibrated to provide adequate information in line with updated international practices. A maintenance programme, including data communication aspects, should be put in place to ensure that no significant lapses occur.

4. DEVELOPMENT OF SEISMIC SOURCE MODELS

GENERAL

4.1. The link between the integrated geological, geophysical, geotechnical and seismological database and the assessment of the seismic hazards is the seismic source model, which should be based on a coherent merging of the individual databases, including due consideration of any available seismotectonic models that may exist or be postulated at the regional scale. The seismic source model constitutes the conceptual and mathematical representation of the physical nature of the seismic sources identified on the basis of the information compiled in the indicated databases and seismotectonic models. One or several seismic source models can be postulated. In the development of such models, all relevant interpretations of the available data should be taken into account, with due consideration of all the uncertainties involved. These models include detailed characterization of the seismic sources and should be developed to be used specifically for the seismic hazard assessment, applying either deterministic or probabilistic approaches.

4.2. The process for developing a seismic source model starts with the integration of the elements of seismological, geophysical, geological and other relevant databases into an integrated database, as recommended in Section 3, to obtain a coherent model (and potential alternative models). This integrated database should also include the available seismotectonic models for the regional scale containing the geographical area of interest and, if necessary, data for beyond the regional scale. These seismotectonic models should also include consideration of the uncertainties embedded either expressly or implicitly in their characterization.

4.3. Using the available data and information included in the integrated database, as well as the interpretations provided by the experts involved, a detailed characterization of all identified and postulated seismic sources should be conducted with the aim of identifying and characterizing in detail all sources of earthquakes that could contribute to the seismic hazard at the site. This source characterization should provide all the necessary characteristics (e.g. location, geometries, potential maximum magnitude, recurrence) of the identified seismic sources.

4.4. The seismogenic structures identified throughout the process of compiling the database might not explain all the observed earthquake activities. This is because seismogenic structures might exist without recognized surface or subsurface manifestations and because different timescales are involved; for example, fault ruptures might have long recurrence intervals with respect to seismological observation periods. Consequently, the seismic source models should consist, to a greater or lesser extent, of two types of seismic source:

(a) Those seismogenic structures that can be identified and characterized using the available database;
(b) Diffuse seismicity (consisting usually, but not always, of small to moderate earthquakes) that might not be attributable to specific seismogenic structures identified in the available database [8].

4.5. The identification and characterization of seismic sources of both types should include assessments of the specific uncertainty involved in each type. Diffuse seismicity poses a particularly complex problem in seismic hazard assessment and will generally involve greater uncertainty because the causative faults of earthquakes are either not well understood or not well characterized with currently available information.

4.6. The development of the seismic source models and the characterization of all parameters of each of their elements should be based primarily on interpretation and evaluation of the available data.

4.7. If the compiled geological, geophysical and seismological data support alternative seismic source models, and the differences in these models cannot be resolved by means of additional investigations within a reasonable time frame, all such models should be taken into consideration in the final hazard evaluation.

4.8. The validity of the proposed seismic source models should be evaluated against existing knowledge and information, for example by comparing long term strain rates predicted by the model against available and reliable geodetic and geological observations.

SEISMOGENIC STRUCTURES (IDENTIFIED SEISMIC SOURCES)

Identification

4.9. All seismogenic structures that might contribute to the seismic hazards at the site should be included in the seismic source models, and uncertainties in the models should be evaluated by sensitivity analysis.

4.10. In the evaluation of fault displacement hazards, special attention and consideration should be given to those seismogenic structures close to the site that have a potential for surface displacement at or near the ground surface (i.e. capable faults; see Section 7). The data collected for this purpose should be evaluated to see whether they are consistent with the data collected for the vibratory seismic hazard analysis. Any inconsistencies should be reconciled if they could adversely affect either analysis.

4.11. The identification of seismogenic structures should consider those geological features for which direct or indirect evidence exists of there having been a seismic source within the current tectonic regime.

4.12. When specific data on a particular geological feature are insufficient for detailed characterization of the feature, a detailed comparison of this feature with other analogous geological features in the region or in similar tectonic regions in the world should be made in terms of age of origin, direction of movement (sense of slip) and history of movement to help determine whether the feature can be considered a seismogenic structure.

Characterization

4.13. For seismogenic structures that have been identified as being relevant to determining the earthquake generated hazards for the site, the associated characteristics of such structures should be determined. The fault geometry (e.g. length, width, depth), orientation (i.e. strike, dip and rake angles), rate of deformation and geological complexity (e.g. segmentation, rupture initiation, secondary faults) should be determined to the extent possible. Determination of these characteristics should be based on an evaluation of all data and information contained in the geological, geophysical, geotechnical and seismological databases.

4.14. Available information about the seismological and geological history of the rupture of a fault or structure (e.g. segmentation, fault length, fault width) should be used to estimate the maximum rupture dimensions and/or displacements. This information, together with magnitude–area scaling relationships, should be used to evaluate the potential maximum magnitude of the seismogenic structure under consideration. Other data that may be used to establish a rheological profile — such as data on heat flow, crustal thickness and strain rate — should also be considered in this estimation.

4.15. In locations where a fault zone comprises multiple fault segments, each fault segment should be taken into account both dependently and independently. The possibility of the multiple fault segments rupturing simultaneously during an earthquake should also be evaluated. To determine the conservative estimate and associated uncertainties of the potential maximum magnitude, the possible scenarios for total fault rupture length should be developed.

4.16. The potential maximum magnitude associated with each seismic source should be specified, and the uncertainty in the potential maximum magnitude should be described by a discrete or continuous probability distribution. For each seismic source, the value of potential maximum magnitude is used as (a) the upper limit of integration in a probabilistic vibratory ground motion hazard calculation to derive the magnitude–frequency relationship and (b) the evaluated scenario magnitude in a deterministic vibratory ground motion hazard analysis. In general, especially for sites in intraplate settings, the largest observed earthquake is a poor and unconservative estimate of potential maximum magnitude. Consideration should be given to the use of appropriate empirical relationships to derive potential maximum magnitude values from controlling or significant faults in the region (e.g. fault geometry, fault rupture mechanism). But if the current fault rupture mechanism cannot be reliably determined, the use of global analogues should be considered, and care should be taken to determine the appropriate seismotectonic analogue. The sensitivity of the resulting hazard with respect to the selection of the potential maximum magnitude values should be tested.

4.17. Other approaches to estimating potential maximum magnitudes on the basis of statistical analysis of the magnitude–frequency relationships for earthquakes associated with a particular structure should also be considered, as appropriate. These approaches assume an association between the structure and all the earthquake data used. In all cases, the results of these methods should be confirmed to be consistent with the available collected data, including palaeoseismological data.

4.18. Irrespective of the approach or combination of approaches used, the determination of the potential maximum magnitude might have significant uncertainty, which should be incorporated into the analysis in a manner consistent with its interpretation in seismological, geological, geophysical and geomorphological data.

4.19. In addition to the potential maximum magnitude, for each seismogenic structure included in the seismic source model, the following characteristics should be determined: (a) the rate of earthquake activity; (b) an appropriate type of magnitude–frequency relationship (e.g. characteristic, exponential); and (c) the uncertainty in this relationship and in its parameters. In the case of the characteristic earthquake occurrence model, the most recent event should be identified as far as possible.

4.20. For those seismic sources for which few earthquakes are registered in the compiled geological and seismological databases, the determination of magnitude–frequency relationships (e.g. the Gutenberg–Richter relationship) may involve a different approach, which may include adopting the coefficients (slope b and intercept a) of the relationship that represents the regional tectonic setting of the seismic source, for example a stable continental tectonic setting. This approach is viable because many studies have shown that the b value of the Gutenberg–Richter relationship varies over a relatively narrow range within a given tectonic setting. Irrespective of the approach used to determine the a and b values of the magnitude–frequency relationship, the uncertainty in those parameters and their correlations should be appropriately assessed and incorporated into the seismic hazard analysis.

ZONES OF DIFFUSE SEISMICITY

Identification

4.21. Zones of diffuse seismicity are those areas in which there is evidence of seismicity that is not attributable to any specific identified seismogenic structures on the basis of the available databases and seismotectonic models. The seismic source model of each zone is developed on the basis that it encompasses an area that possesses similar seismotectonics.

4.22. In the performance of a seismic hazard assessment, knowledge about the depth distribution of the diffuse seismicity (e.g. derived from the seismological, geological and geophysical databases) should be incorporated and the thickness and depth of the seismogenic zone should be properly characterized.

4.23. Significant differences in rates of earthquake occurrence may suggest different tectonic conditions, and they should be considered in defining the boundaries of the zone of diffuse seismicity. Significant differences in focal depths (e.g. crustal versus subcrustal), focal mechanisms, states of stress, tectonic characteristics and Gutenberg–Richter b values may all be used to differentiate between diffuse seismicity zones.

Characterization

4.24. The potential maximum magnitude associated with a zone of diffused seismicity should be evaluated on the basis of seismological data and the seismotectonic characteristics of the diffuse seismicity zone. Comparison with similar world regions for which extensive seismological data are available may be useful, but informed judgement should be used in such an evaluation. Often, the value of potential maximum magnitude obtained will have significant uncertainty owing to the relatively short time period covered by the seismological data with respect to the processes of ongoing deformation. This uncertainty should be appropriately represented in the seismic source model.

4.25. Available information about the seismological and geological history of the seismotectonic structure (e.g. stress regime, strain rate) should be used to estimate the potential maximum magnitude. Other data that may be used to establish a rheological profile — such as data on heat flow, crustal thickness and micro-earthquake distribution — should also be considered in this estimation.

4.26. The potential maximum magnitude associated with each seismic source should be specified, and the uncertainty in potential maximum magnitude should be described by a discrete or continuous probability distribution. For each seismic source, the value of potential maximum magnitude is used as (a) the upper limit of integration in a probabilistic vibratory ground motion hazard calculation to derive the magnitude–frequency relationship and (b) the evaluated scenario magnitude in a deterministic vibratory ground motion hazard analysis. In general, especially for sites in intraplate settings, the largest observed earthquake is a poor and unconservative estimate of potential maximum magnitude. The use of global analogues should be considered, and care should be taken to determine the appropriate seismotectonic analogue. The sensitivity of the resulting hazard with respect to the selection of the potential maximum magnitude values should be tested.

4.27. Other approaches to estimating potential maximum magnitude values on the basis of statistical analysis of the magnitude–frequency relationships for earthquakes associated with a particular structure should also be considered, as appropriate. These approaches assume an association between the structure and all the earthquake data used. In all cases, the results of these methods should be confirmed to be consistent with the available collected data, including palaeoseismological data.

4.28. Irrespective of the approach or combination of approaches used, the determination of the potential maximum magnitude might have significant uncertainty, which should be incorporated into the analysis in a manner consistent with its interpretation in the seismological, geological, geophysical and geomorphological data.

4.29. In addition to the potential maximum magnitude, for each seismogenic structure included in the seismic source model, the following characteristics should be determined: (a) the rate of earthquake activity; (b) an appropriate exponential magnitude–frequency relationship (e.g. the Gutenberg–Richter relationship); and (c) the uncertainty in this relationship and in its parameters.

4.30. For those seismic sources for which few earthquakes are registered in the compiled geological and seismological databases, the determination of magnitude–frequency relationships (e.g. the Gutenberg–Richter relationship) may involve a different approach, which may include adopting the coefficients (slope b and intercept a) of the relationship that represents the regional tectonic setting of the seismic source, for example a stable continental tectonic setting. This approach is viable because many studies have shown that the b value varies over a relatively narrow range within a given tectonic setting. For *a* values, an approach based on strain rates can be used if such data are reliably available from geophysical investigation. However, for many low seismicity areas, *a* values are derived from the regional historical earthquake catalogue (if enough data can be collected), since *a* values are often the most reliable indicator of regional seismicity. Irrespective of the approach used to determine the *a* and *b* values of the magnitude–frequency relationship, the uncertainty in those parameters and their correlations should be appropriately assessed and incorporated into the seismic hazard analysis.

5. METHODS FOR ESTIMATING VIBRATORY GROUND MOTION

GENERAL

5.1. The variability associated with the prediction of vibratory ground motions from future earthquakes is typically one of the largest sources of uncertainty in seismic hazard assessment. Currently available methods for estimating ground motions include GMPEs, which are primarily empirical, and direct simulation methods, which involve physics based scaling to interpolate a smaller amount of data. These alternative methods are described in paras 5.17–5.23. Given the significant epistemic uncertainty currently inherent in ground motion prediction, multiple relationships and/or methodologies should be used. However, the evaluation of ground motion using different methods should be done in a consistent and complementary manner.

5.2. Individual models for the prediction of vibratory ground motions should include both an estimate of the median ground motion amplitude — which, in the case of the commonly adopted log-normal model, is the mean of logarithmic normal distribution — and a measure of the aleatory variability about the mean. The final complete vibratory ground motion model should include an assessment of the epistemic uncertainty in the mean prediction as well as its aleatory variability in the logarithmic scale.

5.3. The definition of the vibratory ground motion intensity used in the ground motion characterization should be consistent with (a) its intended use in subsequent engineering design and probabilistic safety analyses for structures, systems and components of the nuclear installation and (b) the assessment of ground failures such as slope failures and liquefaction. Empirical relationships are typically developed for horizontal response spectral acceleration[6] at 5% of critical damping. Alternative damping levels can be derived using published scaling relationships. Simulation methods typically produce ground motion time histories from which any necessary intensity measure can be derived directly.

5.4. Care should be taken to ensure that the way in which the horizontal components of ground motion are represented in the chosen GMPEs is consistent with their subsequent engineering use in design or fragility analyses. The number of spectral periods characterized should be sufficient to develop smooth spectral shapes (see Section 8).

5.5. The vibratory ground motion should be calculated at a specific location within the soil profile of the nuclear installation site, which is defined as the control point. In some situations, multiple control points may be necessary. The specification of the control point is an important issue in relation to the interface between the vibratory ground motion hazard analysis and the site response analysis. The control point should be clearly defined from the beginning of the project in accordance with the needs of the end user of the evaluation (see Section 10). The control point location could be defined at the free field ground surface, at the outcrop of bedrock or at any other specified depth in the soil profile that is at sufficient depth so that the effects of soil–structure interaction are negligible. The vibratory ground motion specified at the defined control point — to be used as the input for calculating the response of the structures, systems and components of the nuclear installation — should be evaluated and developed through an appropriate site response analysis.

GROUND MOTION PREDICTION EQUATIONS

Selection criteria

5.6. GMPEs specify the median value of vibratory ground motion amplitude on the basis of a limited number of explanatory variables, such as earthquake magnitude, distance from rupture plane (with respect to the site), site conditions and style of faulting. The model may be in the form of an equation or a table. Even for models that are primarily based on empirical data, simulation results are

---

[6] The spectral acceleration is the peak acceleration response of a linear single degree of freedom oscillator as a function of its natural period or frequency and damping ratio when subjected to an acceleration time history.

often used to provide constraints on scaling behaviour for magnitudes, distances or rupture planes that are not well represented in the existing databases. Typically, a set of GMPEs is selected and used in the seismic hazard analysis.

5.7. The selection of the set of appropriate GMPEs should be based on the GMPEs' consistency with the seismotectonic conditions and with the output parameters needed for the seismic hazard assessment (see Section 10). The range of magnitudes, distances and other parameters for which the GMPEs are valid should be checked.

5.8. The selection of candidate GMPEs to be used in the seismic hazard assessment should be based on the following general criteria:

(a) The GMPEs should be current and well established, supported by an adequate quantity of properly processed data.
(b) They should have been determined by appropriate regression analysis to avoid an error in a subjectively fixed coefficient propagating to the other coefficients.
(c) They should be consistent with the types of earthquake and the attenuation characteristics of the site region.
(d) They should match the tectonic environment of the site region as closely as possible.
(e) They should make use of available local ground motion data as much as possible in their definition. If it is necessary to use GMPEs from elsewhere, they should be calibrated by comparing them with as much local strong motion data as possible. If no suitable data are available from the region of interest, a qualitative justification should be provided for why the selected GMPEs are suitable.
(f) They should be consistent with the physical characteristics of the control point location.

5.9. In active tectonic regions, relatively abundant empirical data exist and GMPEs should be developed primarily from those data or from data from similar seismotectonic settings. In areas with lower rates of earthquake activity, where data are much less abundant (e.g. stable continental regions), alternative empirical or semi-empirical methods have been developed for deriving GMPEs. Examples of these methods include the hybrid empirical method and the referenced empirical method, both of which rely on using a GMPE developed for regions where abundant data exist (a host region). In the hybrid empirical method, simple parametric seismological models of the physical properties of the seismic source and diminution of seismic energy with distance are used to adjust the host GMPE to conditions consistent with the site or region of interest (the target conditions). For the referenced empirical method, adjustments[7] should be developed that are based on residuals between the empirical data in the target region and the GMPE model from the host region. This approach requires an adequate amount of empirical data in the target region to perform the necessary residual analysis for the development of the adjustments.

5.10. If adequate data do not exist in the site region to directly develop a reliable suite of GMPEs, then the adjustments described in para. 5.9 should be used to adapt well calibrated GMPEs from other regions so that they satisfy the general criteria in para. 5.8. To avoid the propagation of errors arising from the subjective evaluation of GMPE coefficients, these coefficients should be evaluated on the basis of physics based scaling. If non-ergodic GMPEs are to be used, all coefficients should be properly identified to represent the ground motions for the specific conditions. If ergodic GMPEs are to be used, they will generally be able to capture overall ground motion characteristics with fewer parameters, although the standard deviation might be larger than for non-ergodic GMPEs.

5.11. Aleatory variability should be considered for the GMPEs and derived from the residuals between observed and predicted motions. The residuals might depend on magnitude, distance or the ground motion level itself. At the selected specific site, a detailed site response analysis or a residual investigation using vibratory ground motions recorded at the site should be conducted to reduce the aleatory variability.

5.12. Empirically derived vertical vibratory ground motion should be represented either as a vertical component GMPE or as an empirically derived ratio between vertical and horizontal components of motion. Caution should be exercised in the seismic hazard assessment calculations when the vertical component GMPE is used to predict vertical ground motion, since the characteristics of the vertical component GMPE might differ from those determined for the horizontal case.

5.13. Caution should be exercised in comparing the selected GMPEs with recorded ground motions from small, locally recorded earthquakes. The use of such recordings (e.g. in scaling the selected attenuation relationships) should

---

[7] In high seismicity regions, there are many nuclear installation sites where plenty of strong ground motions have been observed. At these sites, site specific residuals can be determined using the ratio between the observed and predicted motions. The ground motion predicted by GMPEs can be corrected with the site specific residuals. This site specific referenced empirical method is included in the regulatory guidelines of Japan, for example.

be justified by showing that their inferred magnitude and distance scaling properties are appropriate for earthquakes within the ranges of magnitude and distance of greatest concern for the seismic safety of the nuclear installation. Nevertheless, best efforts should be made to reflect those observed data in the selection of the GMPEs.

5.14. When available, macroseismic intensity data may also be used to assign weights to GMPEs or to calibrate the selected GMPEs in those regions where instruments for recording strong motion have not been in operation for a long enough period to provide sufficient amounts of instrumental data. These data may be used at least in a qualitative manner to verify that the GMPEs used to calculate the seismic hazard are representative of the regional ground motion characteristics. The uncertainty in converting from macroseismic intensity data to the desired ground motion intensity metric can be significant; therefore caution should be exercised when adopting these conversions.

Epistemic uncertainties of the technically defensible interpretations

5.15. The appropriate treatment of epistemic uncertainties requires the identification, evaluation and quantification of the range of vibratory ground motions that might occur at a site. Except for regions where a sufficient number of independent, region specific GMPEs have been published, the full quantification of the range of possible ground motions might not be possible using the selection of GMPEs currently available for a specific region. This would require using models from other regions and applying adjustments (as described in paras 5.9 and 5.10) either to render the models more applicable to local conditions or to make the models compatible in terms of predictor variables.

5.16. Several alternative methodologies should be used to represent the centre, body and range of technically defensible interpretations for estimating ground motions at a site from future earthquakes. All methods should begin with the development of a representative suite of GMPEs that satisfy the selection criteria described in para. 5.8. The methodologies to develop weights for individual GMPEs should be based on the degree of confidence in each GMPE and/or approach and on the conformance with existing data. In applying this approach, consideration should be given to developing a representation of the future median ground motions using a suite of GMPEs that are complete (to be extended as much as possible), representative and mutually exclusive.

GROUND MOTION SIMULATION METHODS

5.17. Ground motion simulations provide results that can be used to refine and calibrate empirical GMPEs to directly develop ground motion prediction models and to develop ground motions for specific scenario events. Several simulation methods exist. Any simulation approach used should be carefully validated and calibrated against available recorded data from the region of interest.

5.18. One commonly used approach utilizes a stochastic simulation methodology based on simple parametric models that represent the physical properties of the seismic source and the propagation and attenuation of seismic energy. This methodology can represent the source either as a point source or as a finite fault with rupture that evolves in space and time. This methodology should include the development of region specific parametric models for source, path and site effects, which need to be calibrated with empirical data from the region of interest.

5.19. Alternative ground motion simulation methods use a more direct physical representation of the fault rupture mechanism and the seismic wave propagation. Such physics based methods use fault rupture modelling and path specific wave propagation to estimate ground motions. These procedures might be especially effective in cases where nearby faults contribute significantly to the vibratory ground motion hazard at the site and/or where the existing empirical data are limited (e.g. on the hanging wall of a nearby fault). The physics based methods for fault rupture description fall into two general categories: kinematic and dynamic [7].

5.20. In the kinematic simulation approach, the macro parameters (e.g. rupture area, seismic moment, average stress drop, inhomogeneity of the finite fault) need to be identified and the micro parameters (e.g. distributions of the slip velocity function and rise time) on the finite fault need to be defined. The model parameters cannot be known in advance for future ruptures on a specific fault. Hence, in the simulations these parameter values are represented as random variables with appropriate correlation among them. The specific characteristics of the seismotectonic setting where the site is located should also be given due consideration. A sufficient number of simulations should be conducted to provide a stable estimate of the median ground motions at the site of interest as well as the variability about that median. Kinematic models typically use a stochastic approach to model the high frequency portion of the spectrum as a Green's function. However, the aleatory variability needs to be comparable with that associated with empirical GMPEs, since a potential weakness of such kinematic simulations is their inability to capture variability.

5.21. In the dynamic simulation approach, the state of stress and the friction law properties on the fault need to be defined by, for example, slip weakening friction models that are characterized by the dynamic stress drop, strength excess and critical slip distance distribution on the finite fault. As with the kinematic simulation approach, these properties are unknown for future earthquakes on a specific fault and need to be treated as correlated random variables.

5.22. If recordings of earthquakes exist at or near the site (see para. 3.54), these data should be used either in the calibration of the theoretical Green's function or directly as an empirical Green's function in the range of frequencies with a high signal to noise ratio.

5.23. Potential inhomogeneity of the fault rupture model should be considered such that a high frequency component and a pulse-like signal of the seismic wave could depart from any specific area on the fault. Caution should be exercised to ensure that high frequency and low frequency components are not always generated from the same area on the fault. Furthermore, any available relevant two dimensional or three dimensional heterogeneous crustal structure model that deviates from the assumption of homogeneous horizontal layered models should be considered for more realistic simulation of wave propagation.

6. VIBRATORY GROUND
MOTION HAZARD ANALYSIS

GENERAL

6.1. The approach to be used for assessing the vibratory ground motion hazard at the nuclear installation site should be defined at the beginning of the seismic hazard assessment project. The vibratory ground motion hazard may be evaluated by using probabilistic and/or deterministic methods of seismic hazard analysis (see paras 6.8 and 6.15). The choice of the approach will depend on the national regulatory requirements and the specifications of the end user of the evaluation, which should be documented in the project work plan (see Section 10).

6.2. The vibratory ground motion hazard analysis should use all the elements and parameters of the postulated seismic source models (see Section 4), including the quantified uncertainties. Alternative models proposed by experts in the field of seismic hazard analysis should be formally included in the hazard computation.

6.3. In the vibratory ground motion hazard analysis, both types of uncertainty — aleatory and epistemic — should be considered, irrespective of the approach used.

6.4. Computer codes used in the evaluation of the vibratory ground motion hazard should be able to accommodate the various ground motion prediction and seismic source models defined by the project team for the seismic hazard assessment. It should also be demonstrated that these codes appropriately treat uncertainties.

6.5. Consideration should be given during the hazard analysis to appropriate treatment of the interface between the vibratory ground motion hazard analysis and the site response analysis. This is normally considered by specifying a control point or layer beneath the site where the seismic hazard analysis specifies the ground motion; the site response analysis and/or soil–structure interaction analysis then takes this as its input motion (see SSG-67 [5]). Amplification by decreasing impedance (seismic wave velocity and density) and the attenuation in the subsurface strata should be evaluated for the ground motion estimation close to the control point or layer, except in the case of hard rock sites. Actual subsurface strata are not always horizontally homogeneous, and the inhomogeneity of the subsurface structure — including non-linear effects — may influence the wave propagation. Vertical borehole array measurements of the seismic waves are useful for evaluating the wave propagation characteristics at the site (see paras 6.19–6.24).

6.6. Consideration should be given to the possibility that the ground motion hazard might be influenced by fault rupture driven by human activity (e.g. reservoir loading, fluid injection, fluid withdrawal).

6.7. The design basis may be derived using either a probabilistic or a deterministic approach, while the probabilistic safety assessment of the nuclear installation can only be performed using the results of a probabilistic seismic hazard analysis. Requirements for the use of probabilistic safety assessment for nuclear power plants are established in IAEA Safety Standards Series No. SSR-2/1 (Rev. 1), Safety of Nuclear Power Plants: Design [9]. Requirements for the use of probabilistic safety assessment for research reactors and for nuclear fuel cycle facilities are established in IAEA Safety Standards Series No. SSR-3, Safety of Research Reactors [10], and IAEA Safety Standards Series No. SSR-4, Safety of Nuclear Fuel Cycle Facilities [11].

PROBABILISTIC SEISMIC HAZARD ANALYSIS

6.8. A probabilistic approach should be used when the safety of the nuclear installation against earthquake loading needs to be demonstrated with explicit consideration of the likelihood of occurrence of the relevant seismic hazards (e.g. vibratory ground motion level). Probabilistic approaches consider the rates of recurrence of seismic events for all seismic sources with magnitudes between a bounded minimum magnitude and the estimated potential maximum magnitude. In these cases, the annual frequency of exceedance for different levels of the relevant hazard parameters (e.g. the peak ground acceleration) should be estimated to define an appropriate design basis and/or to perform a seismic probabilistic safety assessment.

6.9. Evaluation of the vibratory ground motion hazard by probabilistic methods should include the following steps:

(1) Selection of the level of effort, resources and details to be applied in the seismic hazard assessment project, considering the safety significance of the nuclear installation, the technical complexity and the uncertainties in the hazard inputs, regulatory requirements and oversight, and the amount of contention within the related scientific community.[8]

(2) Development of a detailed work plan with careful consideration of the experts who will constitute the project team and of the project reviewers who will participate in the independent peer review. If a participatory peer review is envisaged in the project plan, the work plan should enable technical meetings to be held involving experts from the project team and the review team to discuss topics relating to (a) the hazard determination and the availability and quality of the compiled data, (b) alternative interpretations and (c) feedback for implementation of the project. If a participatory peer review is not included in the project plan, its non-inclusion should be justified.

(3) Compilation of the integrated geological, geophysical, geotechnical and seismological database, as recommended in Section 3, and development of the seismic source models for the site region in terms of the defined seismic sources, including uncertainty in their boundaries and dimensions, as recommended in Section 4. A 'zoneless' approach [8] is an alternative

---

[8] The operating organization might also adopt a more resource intensive project as a way of addressing public concern, but this is not a technical judgement, and the merits of such an approach are not considered in this Safety Guide.

scheme to avoid boundary issues, but its application should be adequately justified.

(4) For each seismic source identified in the seismic source models, estimation of the potential maximum magnitude values, evaluation of the rate of earthquake occurrence and derivation of the magnitude–frequency relationship, together with the individual associated uncertainties.

(5) Selection of the appropriate GMPEs for the site region and assessment of the uncertainties in both the mean and the variability of the ground motion as a function of earthquake magnitude and distance from the seismic source to the site. The physics based simulation techniques described in Section 5 are alternative methods for evaluating the ground motion using a sufficient number of calculated time histories to define the centre, body and range of the technically defensible interpretations. The selection and/or adjustment of the GMPEs should be done with consideration of their use in site response analysis (i.e. consideration of step (7) will be necessary).

(6) Establishment of analysis models (e.g. logic trees) and performance of hazard calculations, including sensitivity analysis in a phased approach, starting with a preliminary analysis round and discussion of the preliminary results and ending with a final analysis round that will provide the necessary deliverables defined in accordance with the needs of the end user of the evaluation.

(7) Performance of the site response analysis in cases where site response functions are not included in the ground motion evaluation.

(8) Elaboration, review and confirmation of the final report, including all necessary deliverables.

6.10. The smallest annual frequency of exceedance of interest for which the seismic hazard should be calculated will depend on the eventual use of the probabilistic seismic hazard analysis (i.e. whether for design purposes or for input to a seismic probabilistic safety assessment) and should be indicated in the project plan (see Section 10). This value can be extremely low when it is associated with seismic probabilistic safety assessments, where probabilistic criteria (e.g. core damage frequency, large early release frequency) are low in relation to non-seismic initiators. In such cases, care should be taken to assess the suitability and validity of the database, the seismic source models, the GMPEs and the basis for the expert opinions, since uncertainties associated with these elements can significantly bias the results of the hazard analysis.

6.11. To assist in determining the ground motion characteristics at a site, it is often useful to evaluate the fractional contribution from each seismic source to the total vibratory ground motion hazard by means of a deaggregation process. Such deaggregation may be carried out for a target annual frequency of exceedance, typically the value selected for determining the design basis ground motion. The deaggregation should be performed for at least two ground motion frequency ranges, generally at the low and high ends of the spectrum, which can be used to identify the magnitude–distance pairs that have the largest contribution to the annual frequency of exceedance for the selected ground motion frequency ranges, as well as to provide input for the site response analysis.

6.12. To extrapolate or bound the range of seismic magnitudes represented by the database used in the derivation of the GMPEs, it is necessary to use a corresponding lower limit for the seismic magnitude. The practice has been to combine consideration of this lower limit with an engineering measure linked to a ground motion level associated with a seismic magnitude below which no damage would be incurred by the structures, systems and components important to safety at the nuclear installation. A seismic magnitude value alone is not the best way of representing damage potential. As an alternative to the use of a magnitude measure, the lower bound motion filter may be specified (in terms of an established potential damage parameter, such as the cumulative absolute velocity, the peak ground velocity or the instrumental seismic intensity) in conjunction with a specific value of that parameter for which it can be clearly demonstrated that no significant contribution to damage or risk will occur. The lower bound motion filter should be selected to be consistent with the parameters used in the seismic design and in the fragility analysis as well as in the safety analysis.

6.13. Because of the uncertainties, mainly of an epistemic nature, that are involved at each stage of the hazard assessment process, the assumptions adopted in previous steps and the overall results obtained from the analysis should both be evaluated on the basis of available observations and data from actual seismic events, with due consideration given to the difference between the short period of data availability and the return period usually adopted for seismic design of nuclear installations. This evaluation should be used either to check the consistency of the assumptions and the adequacy of the defined branch of the logic tree or to assign proper weight in the logic tree.

6.14. The results of the vibratory ground motion hazard analysis using a probabilistic approach should be consistent with the typical output shown in the Annex.

DETERMINISTIC SEISMIC HAZARD ANALYSIS

6.15. A deterministic approach can be used as an alternative to the probabilistic approach. Care should be taken to select a conservative scenario of the relevant seismic hazards (e.g. a conservative level for the vibratory ground motion hazard) in line with national practice. In these cases, conservative values of the key hazard parameters should be estimated to define an appropriate design basis for the nuclear installation, corresponding to established safety margins in accordance with application of the concept of defence in depth. The deterministic approach assumes single individual values (i.e. occurring with a probability of 1) for key parameters, leading to a single value for the result, as defined in IAEA Safety Standards Series No. SSG-3, Development and Application of Level 1 Probabilistic Safety Assessment for Nuclear Power Plants [12].

6.16. Deterministic seismic hazard analyses are appropriate for regions where sufficient appropriate data exist for key parameters to identify the scenario earthquake. If this is not the case, the level of statistical uncertainty implied for each parameter can lead to the use of excessively conservative bounding values, which is likely in turn to lead to grossly excessive predictions of seismic hazard levels. The main difference between deterministic analysis and probabilistic analysis is that the former does not employ quantitative statistical methods to explicitly model uncertainties in the parameters; this is an especially important and sometimes dominant consideration in seismic hazard assessments for regions of low seismicity.

6.17. The evaluation of the vibratory ground motion hazard by deterministic methods should include the following steps (the first five steps of this process are similar to those described in para. 6.9 for probabilistic seismic hazard analysis):

(1) Selection of the level of effort, resources and details to be applied in the seismic hazard assessment project, considering the safety significance of the nuclear installation, the technical complexity and uncertainties in the hazard inputs, regulatory requirements and oversight, and the amount of contention within the related scientific community (see also footnote 8).

(2) Development of a detailed work plan with careful consideration of the experts who will constitute the project team and of the project reviewers who will participate in the independent peer review. If a participatory review is envisaged in the project plan, the work plan should enable technical meetings to be held between experts from the project team and the review team to discuss topics relating to (a) the hazard determination and the availability and quality of the compiled data, (b) alternative interpretations

and (c) feedback for implementation of the project. If a participatory peer review is not included in the project plan, its non-inclusion should be justified.

(3) Use of the seismic source models that were compiled as recommended in Section 4, in terms of the defined seismic sources identified on the basis of tectonic characteristics, the rate of earthquake occurrence and the type of magnitude–frequency relationships, including non-Poissonian models if possible.

(4) Evaluation of the potential maximum magnitude for each identified seismic source included in the seismic source models, to be determined with consideration of the uncertainty in potential maximum magnitude values.

(5) Selection of the GMPEs adequate for the region and assessment of the mean and variability of the ground motion, to be obtained as a function of earthquake magnitude and the distance from the seismic source to the site, including the influence of the specific site soil conditions.

(6) Performance of the vibratory ground motion hazard calculation:

(i) For each seismogenic structure, it should be assumed that an earthquake with the potential maximum magnitude occurs at the point of the seismogenic structure closest to the site area of the nuclear installation, with account taken of the physical dimensions of the seismic source. When the seismogenic structure is within the site vicinity and its location and extent cannot be determined with sufficient accuracy, the potential maximum magnitude should be assumed to occur beneath the site.

(ii) For zones of diffuse seismicity that do not include the site, the associated potential maximum magnitude should be assumed to occur at the point of the region closest to the site.

(iii) In a zone of diffuse seismicity that includes the site of the nuclear installation, the potential maximum magnitude should be assumed to occur at some identified specific horizontal and vertical distance from the site. This distance should be based on detailed seismological, geological and geophysical investigations (both onshore and offshore) with the goal of showing the absence of faulting in the site vicinity or, if faults are present, ensuring that they are characterized with the direction, extent, history and/or rate of movements as well as the age of the most recent movement being characterized as older than the established definition for fault capability (see Section 7). This investigation will typically cover an area less than about 10 km. The actual distance used in the GMPEs will depend on the best estimate of the focal depths and on the physical dimensions of the potential fault ruptures for earthquakes expected to occur in the seismotectonic province.

(iv) Several appropriate GMPEs or, in some cases, simulated ground motions based on fault rupture modelling should be used to determine the ground motion that each of the potential maximum magnitude earthquakes would cause at the site, with account taken of the variability of the ground motion.

(v) Ground motion characteristics should be obtained by the deterministic approach, implementing the recommendations provided in para. 5.3.

(7) Consideration of both aleatory and epistemic uncertainties at each step of the deterministic evaluation, to ensure that the conservative procedure described in items (1)–(6) above has covered all the uncertainties involved, while avoiding double counting. This approach should explicitly assess the adequacy of the treatment of uncertainties with respect to the choices that have been made in the different steps (e.g. the assumption that the potential maximum magnitude earthquake would be located at the closest location to the site) to get an appropriate confidence level at the end of the process.

(8) Performance of the site response analysis.

(9) Elaboration, review and confirmation of the final report, including all necessary deliverables.

6.18. If both probabilistic and deterministic assessments are performed, the results from both should be compared. This will enable the deterministic results, including the design basis ground motion, to be calibrated against the probabilistic results, allowing some risk and performance insights to be developed. A further calibration exercise should be performed against the deaggregation analysis to determine the characteristics of the design basis ground motion at the site (see para. 6.11).

SITE RESPONSE ANALYSIS

6.19. Once the vibratory ground motion analysis has been conducted for the selected reference site location and elevation, a site response analysis should be performed that takes into account the detailed and specific geophysical and geotechnical information about the soil profiles in the site area. The aim of the site response analysis is to obtain the vibratory ground motion parameters at the free surface at the top of the soil profile and/or at other locations in the profile, such as the bottom level of the basemat of selected structures and buildings important to safety.

6.20. If the seismic hazard assessment is performed for a new site within which the precise location and layout of the nuclear installation is not yet known (including a lack of information of its foundation characteristics), the site response analysis should be performed at one of the following locations:

(a) The most likely location of the installation within the site area;
(b) A location representative of the general geotechnical characteristics of the site area;
(c) A 'mean' location, that is, an assumed place with mean values of the geotechnical characteristics of the soil profile.

6.21. The site response analysis conducted at this early stage using any of the assumptions in para. 6.20 should be considered as a preliminary site response analysis, needed to define the seismic hazard design basis; it should be followed later by a final site response analysis performed at the finally defined location of the structures of the nuclear installation. It is also possible to defer the site response analysis until the exact location of the structures of the nuclear installation and their foundation parameters are sufficiently well known.

6.22. If the site is an existing site with operating nuclear installations or a site where the specific type of installation is adequately defined in location and layout, the site response analysis should proceed specifically for such installations.

6.23. Two approaches can be taken to properly consider the specific geological and geotechnical soil conditions at a site as part of the estimation of the vibratory ground motion. The first approach is to use GMPEs appropriate for the specific site soil or rock conditions (i.e. using GMPEs that have been developed for subsurface conditions of the type that prevail at the site). The second approach is to conduct site response analyses compatible with the detailed and specific geotechnical and dynamic characteristics of the soil and rock layers at the site area. The decision on which approach to use should therefore be based on the GMPEs used to calculate the seismic vibratory ground motion parameters at the site.

6.24. If the first approach described in para. 6.23 is used, the resulting vibratory ground motion parameters at the free surface of the top of the soil profile may be used directly in defining the seismic hazard design basis for the nuclear installation. If the second approach is used, the following procedure should be applied:

(1) A base case soil profile should be developed with associated soil properties, including parameters to characterize the variability of the soil properties, to ensure consistency with the geophysical and geotechnical databases

compiled as recommended in Section 3, for the full depth from the bedrock outcrop layer to the free surface. The base case soil profile should be defined in terms of the statistical variation of soil properties to accommodate the uncertainties associated with these properties. For each soil layer in the profile, the following parameters should be defined:

(i)   The low strain shear wave velocity (V_S);

(ii) The strain-dependent shear modulus reduction and hysteretic damping properties;

(iii) The soil density;

(iv) The layer thickness;

(v)   For the vertical component, the compressional wave velocity (V_P), if necessary.

(2)   In the case of probabilistic site response analysis, a sufficient number of simulations is necessary to represent the probability distributions of the parameters. In this approach, soil profiles are developed to be consistent with the base case soil profile and to take into account uncertainties associated with the soil properties. The soil profiles generated should be compared with the site specific data to ensure that they are technically justified. The correlation[9] of properties between soil layers in the base case soil profile should also be considered in the development of the low strain simulations. Because soil profiles with quite variable property parameters are modelled as a series of simplified horizontal layers, and because of the oversimplification of the consequences of seismic wave propagation and uncertainties associated with the soil properties, there might be a tendency for resonances of the site response to be overestimated. The probabilistic approach might compensate for this bias by quantitatively modelling the parameter variability.

(3)   Equivalent linear or fully non-linear analyses should be performed for the base case soil profiles as well as for each simulated profile for input ground motions identified either through hazard deaggregation or based on deterministic conservative scenarios. Alternatively, the random vibration theory approach in the frequency domain can be applied by converting between the response spectrum and the Fourier amplitude spectrum to generate mean amplification factors for the site response.

---

[9] Layer to layer correlation is defined as the relation of the probability distribution of a random parameter in one soil layer to the probability distribution of the same parameter in another soil layer (typically, an adjacent layer) within a single profile. The correlation between one parameter of a certain layer and another parameter in another layer may also be specified, if necessary.

(4) The uniform hazard response spectra should be developed at the identified locations of interest (e.g. the control point) for the nuclear installation site and for the annual frequencies of exceedance selected for defining the seismic design basis (e.g. 10^{-4} and 10^{-5} per year). This calculation should take into account the uncertainties in site response (i.e. it should be consistent with the hazard). The final design basis ground motion should be developed with a sufficient safety margin to meet the expectations of a design basis in accordance with the recommendations in SSG-67 [5]. The convolved scenario based ground motion used as input in deterministic approaches should be established as being sufficiently conservative to meet the recommendations.

(5) If the site strata are not horizontally uniform (e.g. valleys, layers with significant inclination), the potential for heterogeneous effects in site response should be examined.

(6) If possible, verification of the results of the site response analysis with any available observed instrumental records should be undertaken, since the site response analysis is complex and its inherent uncertainties might undermine its value in supporting the design of the nuclear installation.

7. EVALUATION OF THE POTENTIAL FOR FAULT DISPLACEMENT AT THE SITE

GENERAL

7.1. In relation to evaluation of fault capability, SSR-1 [1] states (footnote omitted):

"Requirement 15: Evaluation of fault capability

"Geological faults larger than a certain size and within a certain distance of the site and that are significant to safety shall be evaluated to identify whether these faults are to be considered capable faults. For capable faults, potential challenges to the safety of the nuclear installation in terms of ground motion and/or fault displacement hazards shall be evaluated."

. . . .

"5.2. Capable faults shall be identified and evaluated. The evaluation shall consider the fault characteristics in the site vicinity. The methods used and the investigations made shall be sufficiently detailed to support safety related decisions.

"5.3. The potential effect of fault displacement on safety related structures, systems and components shall be evaluated. The evaluation of fault displacement hazards shall include detailed geological mapping of excavations for safety related engineered structures to enable the evaluation of fault capability for the site.

"5.4. A proposed new site shall be considered unsuitable when reliable evidence shows the existence of a capable fault that has the potential to affect the safety of the nuclear installation and which cannot be compensated for by means of a combination of measures for site protection and design features of the nuclear installation. If a capable fault is identified in the site vicinity of an existing nuclear installation, the site shall be deemed unsuitable if the safety of the nuclear installation cannot be demonstrated."

The recommendations provided in this section are aimed at meeting these requirements, with special consideration given to the differences between new sites and existing sites.

7.2. Fault displacement is the relative movement of two sides of a fault at or near the surface, measured in any chosen direction, generated by an earthquake. Primary, or principal, faulting occurs along a main fault rupture plane (or planes) that is the location of release of the energy. Secondary, or distributed, faulting is the rupture that occurs near the principal faulting, possibly on splays of the main fault or on antithetic faults. In other words, displacements could be associated with the causative (i.e. seismogenic) fault or could occur coseismically on secondary faults. Tectonic relative displacements associated with folds (i.e. synclines and anticlines) are also included in the term 'fault displacement'. Fault creep, when demonstrated as such, is considered as a slowly progressing geological hazard that might affect the safety of the nuclear installation but is not seismically induced and is therefore not considered in this Safety Guide.

CAPABLE FAULTS

Definition

7.3. The first question in assessing the potential for fault displacement is whether a fault (buried or outcropping) within the site vicinity and/or within the site area is to be considered capable (i.e. whether the fault has a significant potential for producing displacement at or near the ground surface). The basis for answering such a question should be the proper analysis and interpretation of the data compiled in the integrated database (see Section 3), as incorporated in the seismic source models (see Section 4), together with additional specific data that may be needed for such an assessment.

7.4. On the basis of the geological, geophysical, geodetic and/or seismological data, a fault should be considered capable if the following conditions apply:

(a) If the fault shows evidence of past movement (e.g. significant deformations and/or dislocations) within such a period that it is reasonable to conclude that further movements at or near the surface might occur over the lifetime of the site or the nuclear installation, the fault should be considered capable. In highly active areas, where both seismic and geological data consistently reveal short earthquake recurrence intervals, evidence of past movements in the Upper Pleistocene to the Holocene (i.e. the present) might be appropriate for the assessment of capable faults. In less active areas, it is likely that much longer periods (e.g. the Pliocene to the Holocene (i.e. the present)) are appropriate. In areas where the observed activity is between these two rates (i.e. not as highly active as plate boundaries and not as stable as cratonic zones), the length of the period to be considered should be chosen on a conservative basis (e.g. the Quaternary with possible extension to the Pliocene, depending on the area's tectonic activity level). One way to calibrate the time frame for fault capability would be to check whether the site is in the deformed area of major regional faults. Longer time frames should be used when the site is far away from the potentially deformed areas of these regional structures.

(b) If the capability of a fault cannot be assessed as indicated in (a) because it is not possible to obtain reliable geochronological data by any available method, the fault should be considered capable if it could be structurally linked with a known capable fault (i.e. if a structural relationship with a known capable fault has been demonstrated such that the movement of one fault might cause movement of the other fault at or near the surface).

(c)   If the capability of a fault cannot be assessed as described in (a) and (b) because it is not possible to obtain relevant reliable data by any available method, the fault should be considered capable if the potential maximum magnitude associated with the seismogenic structure, as determined in Section 4, is sufficiently large and the seismic activity is suspected at such a depth (i.e. sufficiently shallow) that it is reasonable to conclude that, in the current tectonic setting of the site area, movement at or near the surface could occur.

7.5.   The period within which evidence of past movement will determine the capability of a fault, as indicated in para. 7.4(a), should be defined at the beginning of the seismic hazard assessment project through a site specific criterion based on the characteristics of the regional tectonic environment and the conditions in the near region and site vicinity. This criterion for assessing fault capability should be established by or agreed with the regulatory body.

Investigations necessary to determine capability

7.6.   Sufficient surface and subsurface related data should be obtained from the investigations in the regional, near regional, site vicinity and site areas (see Section 3) to demonstrate the absence of faulting at or near the site or, if faults are present, to describe the direction, extent, history and rate of movement of these faults as well as the age of the most recent movement.

7.7.   When surface faulting is known or suspected to be present, investigations should be conducted at the site vicinity scale and should include very detailed geological and geomorphological mapping, topographical analyses, geophysical surveys (including geodetic measurements, if necessary), trenching, boreholes, age dating of sediments or faulted rock, local seismological investigations, and any other appropriate and state of the art techniques (e.g. remote sensing methods) to ascertain the amount and age of previous displacements or deformations.

7.8.   Consideration should be given to the possibility that faults that have not shown recent near surface movement might be reactivated by human activity (e.g. reservoir loading, fluid injection, fluid withdrawal).

7.9.   Investigations of a capable fault should be sufficient to enable a confident decision to be made regarding whether the fault can be screened out as a credible hazard to nuclear safety or, if the hazard is judged to be credible, to provide sufficient quantitative information to the subsequent site evaluation, design and safety analysis process in accordance with para. 10.3. The capable fault investigations should also link to those investigations undertaken for vibratory ground motion analysis and should be consistent with them. While the specific needs of both analyses are somewhat inconsistent in terms of data needs and outputs, the documented narrative that reports on these analyses should recognize that both hazards derive from the same tectonic structures in the region.

CAPABLE FAULT ISSUES FOR PROPOSED NEW SITES

7.10. In the selection and evaluation stages of a proposed new site for a nuclear installation, if reliable evidence is collected demonstrating the existence of a capable fault with potential for seismogenic (i.e. primary) fault displacement within the site vicinity, or within the site area, and its effects cannot be compensated for by proven design or engineering protective measures, this issue should be treated as an exclusionary attribute (see para. 3.8 of IAEA Safety Standards Series No. SSG-35, Site Survey and Site Selection for Nuclear Installations [13]) and an alternative site should be considered.

7.11. In the selection and evaluation stages of a proposed new site, if reliable evidence is collected demonstrating the existence within the site vicinity of a secondary fault associated with a capable fault located outside the site vicinity, the secondary fault may be treated as a discretionary attribute (see para. 3.8 of SSG-35 [13]). However, if reliable evidence shows that this secondary fault can be traced to or could extend to the site area, and its effects cannot be compensated for by proven design or engineering protective measures, the existence of this secondary fault should be treated as an exclusionary attribute and an alternative site should be considered. If there is insufficient evidence or data to differentiate between primary and secondary faults, a conservative approach should be applied and such faults should be identified and characterized as capable faults.

CAPABLE FAULT ISSUES FOR SITES WITH EXISTING NUCLEAR INSTALLATIONS

7.12. In general, because of the extensive site investigation programme required for a nuclear installation, the situation will not arise in which further consideration needs to be given to the potential for fault displacement at the site of an existing nuclear installation. However, information might come to light later that there is potentially a capable fault in the site vicinity that requires assessment of the potential for fault displacement. Therefore, for existing nuclear installations for which a seismic safety evaluation programme is conducted (see IAEA Safety

Standards Series No. NS-G-2.13, Evaluation of Seismic Safety for Existing Nuclear Installations [14]), the programme should include the assessment of the fault displacement potential based on the information available from the original site selection and evaluation stages as well as updated information and current techniques and criteria, ensuring proper interpretation of all newly available data.

7.13. If a new nuclear installation is to be built on a site where there is one or more existing nuclear installations and information comes to light that there is potentially a capable fault in the site vicinity, the approach for the new installation should be as recommended in paras 7.10 and 7.11.

7.14. If there is potentially a capable fault within the site vicinity and/or site area, the fault should be characterized to establish whether it could potentially approach the nuclear installation and subsequently cause surface displacement that affects items important to safety. This evaluation should be based on the characteristics of the fault, such as its sense of slip and geometry (i.e. length, width, depth and coordinates, including strike, dip and rake angles). For structurally related (i.e. secondary) faults, the evaluation should be also based on its relationship with the causative fault. The evaluation should use validated empirical and/or theoretical models in a conservative way, including due consideration of related uncertainties, both epistemic and aleatory.

7.15. If no sufficient basis is provided to decide conclusively that the fault is not capable, and if the identified fault has the potential to affect the foundations of items important to safety of the nuclear installation, then, using all the available data compiled as recommended in Section 3, probabilistic methods should be used to estimate the annual frequency of exceedance of various amounts of displacement at or near the surface.

7.16. In the probabilistic fault displacement hazard analysis, the following two types of possible displacement should be considered, with careful and appropriate treatment of the uncertainties involved (both epistemic and aleatory):

(a) Primary or principal displacement that occurs along a main plane (or planes) that is the locus of release of seismic energy.
(b) Secondary or distributed displacement that occurs in the vicinity of the principal displacement, possibly on splays of the main fault or antithetic faults. In some cases, triggered slip[10] has been considered a form of secondary or distributed displacement.

---

[10] A triggered slip is a remote triggering of slip along a fault from a distant earthquake.

The fault displacement is generally characterized as a three dimensional displacement vector that is resolved into components of slip along the fault trace and along the fault dip, with the resulting amplitude equal to the total evaluated slip (for a given annual frequency of exceedance and for a given fractile of hazard).

7.17. The annual frequency of exceedance corresponding to various amounts of displacement at or near the surface should be determined at the foundation points, in accordance with the specific layout of the foundations of the structures, systems and components important to safety of the nuclear installation. The most up to date and reliable methods of probabilistic assessment should be applied. These include empirical relationships and/or engineering models (e.g. finite element analysis, Coulomb static stress transfer models) that are compatible with the faulting type and site area specific geological setting and use all available data.

7.18. The range of annual frequencies of exceedance for which fault displacements are calculated should be compatible with the safety significance of the nuclear installation. This will enable a fault displacement hazard curve to be constructed over the frequency range of relevance to nuclear safety for the installation. The response of the installation to these displacements can be evaluated to determine its capacity to withstand probabilistic fault displacement hazard (i.e. the probability of failure as a function of fault displacement). From both the hazard curve and the failure probability function, the frequency of failure due to fault displacement hazard can in principle be calculated, and this could be compared with relevant regulatory safety goals, such as large early release frequency, that apply to the installation. On the basis of this information, a judgement could be made as to whether the installation meets the intent of Requirement 20 and para. 5.27 of SSR-2/1 (Rev. 1) [9] in terms of the 'practical elimination' of event sequences that could lead to an early radioactive release or a large radioactive release. (See also para. 2.11 of SSR-2/1 (Rev. 1) [9], para. 6.8 of SSR-3 [10] and para. 6.7 of SSR-4 [11].)

8. PARAMETERS RELATING TO VIBRATORY GROUND MOTION HAZARDS, FAULT DISPLACEMENT HAZARDS AND OTHER HAZARDS ASSOCIATED WITH EARTHQUAKES

VIBRATORY GROUND MOTION HAZARDS

Parameters and control point

8.1. Irrespective of the method applied (i.e. a probabilistic approach, a deterministic approach, or both), the vibratory ground motion hazards at the site should be defined by means of appropriate parameters, such as spectral representations and time histories.

8.2. In principle, the vibratory ground motion parameters should be defined at the control point established by the needs of the end user of the evaluation (see Section 10). Usually, the control point is defined at free field conditions (i.e. at the ground surface, at key embedment depths or at bedrock level). In cases where surface soil layers will be completely removed, the parameters should be defined at the level of the outcrop that will exist after removal. Consideration should be given to appropriate treatment of the interface between the defined reference ground motion and the site response analysis.

Site response analysis

8.3. The site response analysis, performed as recommended in paras 6.19–6.24, provides the vibratory ground motion parameters at locations relevant for the design and safety assessment of the nuclear installation (e.g. at the free field ground surface, at foundation level).

Spectral representations

8.4. The vibratory ground motion hazard, calculated as recommended in Section 6, should be characterized by response spectra in horizontal and vertical components at the control point.

Uniform hazard response spectra

8.5. A uniform hazard response spectrum is developed by selecting the values of the response spectral ordinates that correspond to the annual frequencies of exceedance of interest from the seismic hazard curves for individual frequencies or periods. One or more uniform hazard response spectra may be developed from the results of the probabilistic seismic hazard analysis and any subsequent site response analyses that have been performed.

Response spectra based on scenario earthquakes

8.6. In deterministic seismic hazard analyses, as well as after the deaggregation process in the probabilistic seismic hazard analyses, scenario earthquakes should be used to realistically represent the frequency content of ground motions. Scenario earthquakes from the deaggregation process for the results of probabilistic seismic hazard analyses should be associated with annual frequency of exceedance values.

Standardized response spectra

8.7. A standardized response spectrum with a smooth shape is used for engineering design purposes and to take into account the contribution of multiple seismic sources represented by an envelope incorporating adequate low frequency and high frequency ground motion inputs. The prescribed shape of the standardized response spectrum is obtained from various response spectra based on earthquake records and engineering considerations. This standardized response spectrum should be scaled to envelop the median ground motion levels in a wide frequency range.

8.8. It is possible to have low to moderate magnitude near field earthquakes that have a relatively rich high frequency content and short duration with a high peak ground acceleration. The use of the peak ground acceleration from this type of earthquake to scale a broad banded standardized response spectrum could lead to an unrealistic shape for the standardized response spectra. In such a case, multiple response spectra should be used for design purposes to properly reflect the different types of seismic source.

Time histories

8.9. Time histories should satisfactorily reflect all the prescribed ground motion parameters as embodied in the response spectra or other spectral representation, with the addition of other parameters such as duration, phase and coherence. The number of time histories to be used in the detailed analyses and the procedure to be used in generating those time histories will depend on the type of analysis to be performed and should be specified by the end user of the evaluation (see Section 10) on the basis of the different types of engineering analysis to be conducted in the design or safety assessment stages.

8.10. Significant progress has been made in ground motion simulation based on fault rupture modelling with wave propagation paths and site effects (e.g. by use of empirical Green's function methods). Ground motions obtained in this way for regions for which pertinent parameters are available can be employed to complement the more traditional methods. Time histories should be applied carefully, especially when developed for soils that are expected to respond non-linearly.

8.11. In using response spectra to develop design time histories, it should be ensured that the time histories include the appropriate energy content represented by the design ground motions. This could be done by calculating the corresponding power spectral density functions.

Ground motion duration

8.12. The duration of the vibratory ground motion is determined by many factors, including the size of the fault rupture (generally characterized by magnitude), crustal parameters along the propagation path (generally characterized by distance) and conditions beneath the site (e.g. the presence of a significant sedimentary basin). A consistent definition of duration should be used throughout the evaluation. Common definitions of duration include the following:

(a) The time interval between the onset of ground motion and the time at which the acceleration has declined to 5% of its peak value;
(b) The time interval between the 5th percentile and the 95th percentile (75th percentile for high noise records) of the integral of the mean square value of the acceleration;
(c) The time interval for which the acceleration exceeds 5% of the acceleration due to gravity.

8.13. In determining an appropriate duration for the time histories, due weighting should be given to any empirical evidence provided by the regional database. For some sites, relatively low amplitude motions from distant, large earthquakes might pose a liquefaction hazard. In such situations, time histories used for liquefaction should include such low amplitude time histories over an appropriate duration.

Vertical ground motion

8.14. Vertical vibratory ground motions (response spectra and time histories) should be developed using the same methods used for developing horizontal vibratory ground motions. If vertical attenuation relationships are not available, it may be reasonable to assume a ratio between vertical and horizontal ground motion that is prescribed by current best practice. However, caution should be exercised if using GMPEs defined separately for each component (see para. 5.12).

Ground motion for base isolated structures, buried structures and liquid reservoirs

8.15. The methodology for deriving the design ground motions has been developed for installation structures with conventional foundations. For structures that use base isolation systems for protection of the installation against earthquake generated vibratory ground motions, additional considerations may be necessary, including the careful review of worldwide experience in relation to specific performance and design criteria, as well as corresponding regulatory requirements. Of most concern are effects of a long predominant period that might cause excessive residual displacements in the elements of the base isolation system. For structures of the installation for which a base isolation system is envisaged, time histories should be examined and, if necessary, modified to take these effects of the long predominant period (and potentially long duration) into account. The evaluation should consider surface wave influences due to thick sediments.

8.16. For buried structures such as ducts and piping, appropriate response spectra and time histories should be developed to be consistent with parameters used in the structural design.

8.17. An appropriate representation of the vibratory ground motion should be developed when the project plan calls for consideration of sloshing effects in liquid reservoirs (e.g. pools, ponds, tanks with independent foundations).

FAULT DISPLACEMENT HAZARDS

8.18. For existing nuclear installations for which a fault displacement hazard analysis was performed in accordance with paras 7.12–7.18, the surface fault displacement associated with each capable fault that can produce surface faulting in the site area should be determined. These values should correspond to the acceptable value of the annual frequencies of exceedance specified in accordance with the safety requirements established in SSR-1 [1], as well as specified in the project plan. Empirical fault displacement models have a larger uncertainty than vibratory ground motion models due to there being fewer data available, and this should be taken into consideration accordingly.

OTHER HAZARDS ASSOCIATED WITH EARTHQUAKES

8.19. The results of the seismic hazard analysis should also be used in the assessment of other hazards associated with earthquakes that might be significant for the safety of the nuclear installation. These hazards include tsunamis, soil liquefaction, slope instability, subsidence, karstic features and collapse of subsurface cavities, and the failure of water retaining structures that might be triggered either by ground motion or by surface faulting. A thorough assessment should be carried out to determine the level of seismic hazard or the supporting models appropriate for the associated hazard under consideration.

Tsunamis

8.20. For coastal sites, the potential for tsunamis should be carefully evaluated in the framework of hydrological hazards (see SSG-18 [4]). Tsunamis can be generated by earthquakes that cause tectonic deformation of the seabed or submarine landslides. For tectonically generated tsunamis, the region of investigation might be very large, extending to several thousands of kilometres in radius. The investigation should concentrate on those seismic sources with the potential to generate significant vertical displacement of the seabed, since it is this motion that is most likely to cause a tsunami.

8.21. For a tsunami hazard associated with a near regional submarine landslide, the seismic hazard appropriate for triggering the landslide should be determined consistently with the hazard level associated with the nuclear installation site.

8.22. For evaluating the fault related tsunami hazards, the coastal subsidence and uplift should be estimated. A study of palaeo-tsunamis should be conducted within the near region to understand the history of tsunamis on the coast. This assessment may be part of the seismic hazard assessment or the tsunami hazard assessment, but — in any case — the assessments should be coordinated.

Liquefaction potential

8.23. Non-cohesive soils in loosely deposited conditions below the water table are susceptible to liquefaction; if this is the case, the bearing capacity (strength and stiffness) of the soil are reduced when subjected to vibratory ground motions. Therefore, careful geotechnical investigations should be carried out in the site area to assess the liquefaction potential of the soil, including non-cohesive backfill materials, which might affect the safety of the structures, systems and components of the nuclear installation.

8.24. For soils susceptible to liquefaction, detailed information on the modelled soil profile is needed, and it should be obtained as described in paras 3.16 and 3.17 of NS-G-3.6 [3]. To assess the liquefaction potential using any of the three methods described in paras 3.18–3.25 of NS-G-3.6 [3], the specific characteristics of the seismic design basis, or of the seismic hazards at the site, should be provided. Therefore, if an empirical approach is used for assessing the liquefaction potential (see para. 3.19 of NS-G-3.6 [3]), the earthquake magnitude values for different design conditions should be properly defined using the corresponding information and data used for the seismic hazard analysis. If an analytical approach is used (see paras 3.20–3.25 of NS-G-3.6 [3]), earthquake features should be properly identified in relation to the appropriate selection of the time histories to define the number of cycles of stress and the adequate input motions for non-linear stress analysis. In any case, close coordination should be established with the geotechnical engineering experts performing the liquefaction analysis and foundation design. The selection of a site with potential massive liquefaction should be avoided (see annex I to SSG-35 [13]).

Slope instability

8.25. The stability of natural and human-built slopes located in the site area and site vicinity that can be affected by the vibratory ground motions should be investigated, since landslides could seriously affect structures, systems and components important to safety. The stability of slopes should be evaluated using appropriate parameters of the vibratory ground motions obtained from the seismic hazard analysis at the site. As described in para. 5.5 of NS-G-3.6 [3], the peak ground acceleration of the seismic design basis is typically the parameter used for estimating the inertial loads, although in some cases a more refined dynamic analysis may be necessary.

Collapse due to cavities and subsidence phenomena

8.26. The potential for complex subsurface conditions should be investigated, as recommended in paras 2.35–2.47 of NS-G-3.6 [3]. Such conditions at the site area could have serious implications for the integrity of the foundations of items important to safety of the nuclear installation. When performing the seismic hazard assessment for a nuclear installation site, the prediction, detection and evaluation of subsurface conditions should use data and methods adequate for such purposes. As cavities can preferentially develop along seismogenic structures, the potential for coseismic movement of seismogenic structures should be investigated.

Failure of water retaining structures (dam break)

8.27. The potential failure of water retaining structures located upstream of the site area due to a seismic event should be investigated, with consideration of the consequential flooding hazards that might affect the safety of the nuclear installation. Therefore, the seismic design basis, including the seismic hazard and the performance and safety criteria, adopted for such structures should be obtained from the authorities and organizations responsible for such structures. This information should be properly analysed, including specific characteristics (e.g. the water mass controlled or retained by the dams), to ensure the safety of the nuclear installation at the site or to implement adequate site related mitigatory measures.

8.28. Consideration should be given to the possible existence of several dams in the upper stream region for which a domino effect could occur. Hydrodynamic impacts should be considered and should be based on the inundation level as well as the velocity of the water flow. A landslide might produce mud flows, floating debris and temporary landslide dams, and the potential for these dams to break is highly uncertain.

8.29. If all the seismogenic structures that might affect the water retaining structures considered are within the region of investigation for the seismic hazard analysis of the nuclear installation, then the same seismic source characterization used for ground motion models and fault displacement models for the nuclear installation should be used in the seismic hazard assessment of the water retaining structures. If this is not the case, seismic sources common to both the nuclear installation and the water retaining structures should be modelled, taking into account the attributes used in the seismic hazard analysis of the nuclear installation. In any case, close coordination should be established with the hydrological engineering experts performing the dam break analysis and the design of protection against flooding.

Volcano related phenomena

8.30. Earthquakes and related hazards are phenomena associated with volcanic events, as indicated in table 1 of IAEA Safety Standards Series No. SSG-21, Volcanic Hazards in Site Evaluation for Nuclear Installations [15]. Earthquakes generated by volcanic activity are typically smaller than tectonic earthquakes. If a capable fault is identified in the vicinity of an active volcano, both the seismic hazards and the volcanic hazards should be taken into account, since earthquakes might occur on the capable fault preceding, accompanying or following the volcanic eruption as a result of the mutual influence of tectonic movement and magma intrusion. In addition, the identification of aligned volcanic vents in a well defined local area might indicate the presence of a tectonic fault or possibly a capable fault.

9. EVALUATION OF SEISMIC HAZARDS FOR NUCLEAR INSTALLATIONS OTHER THAN NUCLEAR POWER PLANTS

GENERAL

9.1. The evaluation of seismic hazards for nuclear installations other than nuclear power plants should be commensurate with the complexity of such installations, with the potential radiological hazards and with the hazards due to other materials present on the site.

9.2. The recommended method for applying the graded approach is to start with attributes relating to nuclear power plants and, if possible, to commensurately adjust these for installations with which lesser radiological consequences are associated. If this approach is not practicable for a nuclear installation other than a nuclear power plant, then the recommendations relating to nuclear power plants should be applied.

SCREENING PROCESS

9.3. Prior to adopting a graded approach, a conservative screening process should be applied in which it is assumed that the entire radioactive inventory of the installation is released by the potential seismically initiated accident. If the potential result of such a radioactive release is that unacceptable consequences would not be likely — for workers or the public (i.e. doses to workers and to the public would be below the dose limits established by the regulatory body) or for the environment — and if no other specific requirements are imposed by the regulatory body for such an installation, the installation may be excluded from the requirement to undertake a full seismic hazard assessment. If, even after such a result is reached, some degree of seismic hazard assessment is considered necessary, national seismic codes for hazardous and/or industrial facilities should be used.

9.4. If the results of the conservative screening process show that the potential consequences of such a release would be unacceptable, a seismic hazard assessment of the installation should be carried out, starting from the recommendations relevant to nuclear power plants.

9.5. The conservative screening process described in para. 9.3 should consider the likelihood that a seismic event will result in an event with radiological consequences. This likelihood will highly depend on the following factors relating to the characteristics of the nuclear installation (e.g. its purpose, layout, design, construction and operation):

(a) The amount, type and status of the radioactive inventory at the site (e.g. whether solid, liquid and/or gaseous; whether the radioactive material is being processed or only stored);
(b) The intrinsic hazard associated with the physical processes (e.g. nuclear chain reactions) and chemical processes (e.g. for fuel processing purposes) that take place at the installation;
(c) The thermal power of the nuclear installation, if applicable;
(d) The configuration of the installation for different kinds of activity;
(e) The distribution of radioactive sources in the installation (e.g. for research reactors, most of the radioactive inventory will be in the reactor core and the fuel storage pool, whereas for fuel processing and storage facilities it might be distributed throughout the installation);
(f) The changing nature of the configuration and layout of installations designed for experiments (such activities have an associated intrinsic unpredictability);
(g) The need for active safety systems and/or operator actions for the prevention of accidents and for mitigation of the consequences of accidents, and the characteristics of engineered safety features for the prevention of accidents

and for mitigation of the consequences of accidents (e.g. the containment and containment systems);

(h) The characteristics of the structures of the nuclear installations and the means of confinement of radioactive material;

(i) The characteristics of the processes or of the engineering features that might show a cliff edge effect in the event of an accident;

(j) The characteristics of the site that are relevant to the consequences of the dispersion of radioactive material to the atmosphere and the hydrosphere (e.g. size and demographics of the region);

(k) The potential for on-site and off-site contamination.

9.6. Depending on the criteria applied by the regulatory body, some or all of the factors in para. 9.5 should be considered when applying the conservative screening process. For example, the fuel damage, the radioactive release or the doses to workers and the public could be factors that warrant special consideration.

9.7. The application of the graded approach should be based on the following information:

(a) The existing safety analysis report for the installation, which should be the primary source of information;

(b) The results of a probabilistic safety assessment, if one has been performed;

(c) The characteristics specified in para. 9.5.

CATEGORIZATION PROCESS

9.8. If the conservative screening process indicates that a seismic hazard assessment of the installation is to be carried out (see para. 9.5), a process for categorizing the installation should be undertaken. This categorization may be performed at the design stage or later. If the categorization has been performed, the assumptions on which it was based should be reviewed and verified. In general, the criteria for categorization should be based on the radiological consequences of a radioactive release from the installation, ranging from very low to potentially severe consequences. As an alternative, the categorization may consider the radiological consequences within the installation itself, within the site of the installation, and for the public and the environment.

9.9. Three or more categories may be defined on the basis of national practice and criteria, as well as the information described in para. 9.7. As an example, the following categories may be defined:

(a) The lowest hazard category, which includes those nuclear installations for which national building codes for conventional installations (e.g. essential facilities such as hospitals) or for hazardous facilities (e.g. petrochemical or chemical plants) should be applied as a minimum;

(b) The highest hazard category, which includes installations for which standards and codes for nuclear power plants should be applied;

(c) There is often at least one intermediate category between (a) and (b), corresponding to a hazardous installation for which, at a minimum, codes dedicated to hazardous facilities should be applied.

VIBRATORY GROUND MOTION HAZARD ANALYSIS AND ASSOCIATED ASPECTS

Vibratory ground motion hazard analysis

9.10. The vibratory ground motion hazard analysis for installations categorized as recommended in paras 9.8 and 9.9 should be performed in accordance with the following:

(a) For the least hazardous installations, the input ground motion for the design may be taken from national building codes and maps.

(b) For installations in the highest hazard category, methodologies for seismic hazard assessment as described in Sections 3–8 of this Safety Guide (i.e. recommendations applicable to nuclear power plants) should be used.

(c) For installations categorized in the intermediate hazard category, the following approach might be applicable:

   (i) If the seismic hazard assessment is typically performed using methods similar to those described in this Safety Guide, a lower input ground motion than that evaluated for (b) may be adopted for designing these installations, in accordance with the safety requirements for the installation.

   (ii) If the database and the methods recommended in this Safety Guide are found to be disproportionately complex, time consuming and demanding for the nuclear installation in question, simplified methods for seismic hazard assessment (that are based on a more restricted data set) may be used. In such cases, the input ground motion finally adopted for designing the installation should be commensurate with the reduced database and the simplification of the methods, with account taken of the fact that both factors tend to increase uncertainties.

9.11. The design basis ground motion levels for nuclear installations other than nuclear power plants should be decided in the context of the approach to hazard assessment recommended in para. 9.10.

9.12. The recommendations relating to seismic instrumentation installed on the site (see paras 3.54–3.59) should be applied in a manner commensurate with the category of the installation, as defined in para. 9.9.

Geological and geotechnical aspects associated with seismic hazards

9.13. With regard to the geological and geotechnical aspects associated with seismic hazards, the same considerations used for nuclear power plants should apply to other types of nuclear installation. If reliable evidence demonstrates that fault displacement phenomena arising from these aspects could occur within the site vicinity and/or site area, a detailed and specific fault displacement assessment should be conducted. The site may still be considered suitable on the basis of specific established suitability criteria, and design bases should be established to ensure the safety of the nuclear installation through design, construction and operation measures.

10. APPLICATION OF THE MANAGEMENT SYSTEM

ASPECTS OF PROJECT MANAGEMENT

10.1. A management system, to be established, applied and maintained as required by IAEA Safety Standards Series No. GSR Part 2, Leadership and Management for Safety [16], should be implemented for the activities performed for the seismic hazard assessment of the site.

10.2. A project work plan should be established that, at a minimum, addresses the following topics:

(a) The objectives and scope of the project;
(b) Applicable regulations and standards;

(c)   Organization of the roles and responsibilities for management of the project;

(d)   Work breakdown, processes and tasks, and schedule and milestones;

(e)   Interfaces among the different tasks (e.g. field tasks, laboratory tests, analysis) and disciplines (e.g. earth sciences, engineering) involved, with all necessary inputs and outputs;

(f)   Project deliverables and reporting.

10.3. The project scope should identify all the hazards generated by earthquakes relevant to the safety of the nuclear installation that will be investigated within the framework of the project. This Safety Guide addresses individual hazards associated with earthquakes. Depending on the objectives of the project, some or all of these hazards may be considered in the scope. If some of the hazards are considered to be out of scope because it is believed that they are not relevant to the site, a screening process should be applied to demonstrate and document that this is the case.

10.4. The project work plan should include a description of all requirements that are relevant for the project, including applicable regulatory requirements in relation to all the hazards considered to be within the project scope. The applicability of the set of regulatory requirements should be reviewed by the regulatory body prior to conducting the seismic hazard analysis.

10.5. All approaches and methodologies that reference lower tier regulations (e.g. regulatory guidance documents, industry codes and standards) should be clearly identified and described. If procedures for experts' interaction are used to better capture epistemic uncertainties, the sophistication and complexity of these approaches should be chosen by the operating organization on the basis of the project requirements. The details of the approaches and methodologies to be used should be clearly stated in the project work plan. These details should include the functions of the various experts involved in the project (e.g. proponent, resource expert, technical integrator, review panel member) and their responsibilities with regard to the project.

10.6. At least the following generic management system processes should be applied to ensure the quality of the project: document control, control of products, controls for measuring and testing equipment, control of records, control of analyses, purchasing (procurement), verification and validation of software, audits (e.g. self-assessment, independent assessments and review), control of non-conformances, corrective actions, preventive actions, and management of human resources [17]. Processes covering field investigations, laboratory testing, data collection, and analysis and evaluation of observed data should be applied.

Communication processes for interaction among the experts involved in the project should be also applied.

10.7. The project work plan should ensure that there is adequate provision, in the resources and in the schedule, for collecting new data that might be important for the conduct of the seismic hazard assessment or for responding to requests by experts, including provision for balancing potentially conflicting project needs.

10.8. To make the evaluation traceable and transparent to its end user (e.g. peer reviewers, the operating organization, the regulatory body, the designers, the vendors, and the contractors and subcontractors of the operating organization), the documentation for the seismic hazard assessment should provide a description of all elements of the project, including the following information:

(a) A description of the participants in the evaluation and their roles;
(b) Background material that contains the analysis documentation, including raw data and processed data;
(c) A description of the computer software used, and input and output files;
(d) Reference documents;
(e) All documents supporting the treatment of uncertainties, expert opinion and related discussions;
(f) Results of intermediate calculations and sensitivity studies.

This documentation should be maintained in an accessible, usable and auditable form by the operating organization.

10.9. The documentation and references should identify all sources of information used in the seismic hazard assessment, including information on where to find important citations that might be difficult to obtain. Unpublished data that are used in the assessment should be included in the documentation in an appropriately accessible and usable form. Documentation or references that are readily available elsewhere should be cited where appropriate.

10.10. The documentation for the seismic hazard assessment should identify the computer software that was used. This should include the computer programs used in the processing of data (e.g. the earthquake catalogue) and the computer programs used to perform calculations for the seismic hazard.

10.11. Owing to the variety of investigations carried out (e.g. field investigations, laboratory tests, calculations) and the need for expert judgement in the decision making process, technical procedures specific to the project should be developed to guide and facilitate the execution and verification of these processes.

ENGINEERING USES AND OUTPUT SPECIFICATION

10.12. A seismic hazard assessment is usually conducted for the purposes of seismic design and/or seismic probabilistic safety assessment of the nuclear installation. Therefore, from the beginning, the work plan for the seismic hazard assessment should identify the intended engineering uses and objectives of the assessment and should specify the necessary outputs (i.e. all the results necessary for the intended engineering uses and objectives of the assessment).

10.13. To the extent possible, the output specification for the seismic hazard analysis should be comprehensive. The output specification may be updated, as necessary, to accommodate additional results and/or to reduce the scope of the results. The output specification should consider the following elements:

(a) Ground motion parameters: Specified ground motion parameters should be sufficient to produce the necessary results and any additional outputs needed for engineering uses (see the Annex for typical outputs of a probabilistic seismic hazard analysis for assessing the vibratory ground motion parameters).
(b) Predominant frequencies: The range and density of specified predominant frequencies for the uniform hazard response spectra should be sufficient to adequately represent the input for all structures, systems and components important to safety.
(c) Damping: Specified damping values should be sufficient to adequately represent input for analysing the responses of all structures, systems and components important to safety and the effects on such items.
(d) Ground motion components: The output of both vertical and horizontal motions should be specified.
(e) The reference subsurface rock site condition: For site response analysis, the output should be specified on the rock conditions at the site (usually to a depth significantly greater than 30 m, corresponding to a specified value of the shear wave velocity consistent with firm rock). The results of site response analysis should correspond to this reference condition.
(f) Control points: The output specification should specify the control points (e.g. depths at the site) for which the results of a near surface vibratory ground motion hazard analysis are obtained. Usually, the control points are set at the ground surface and at key embedment depths (e.g. foundation

levels) for structures and components. The specified control points should be sufficient to develop adequate inputs for soil–structure interaction analyses.

10.14. In any seismic hazard assessment, there is a need to consider a lower bound magnitude owing to constraints in the seismological database (see para. 6.12). Therefore, in addition to the output specification for anticipated engineering uses, the project plan should specify the following additional parameters relating to engineering validity or the utility of the seismic hazard analysis:

(a) Lower bound motion filter: Use of a lower bound motion is needed for practical computation purposes in the seismic hazard analysis and the lower bound motion should be selected to include all events with potential radiological consequences. The filter for the lower bound motion should be selected to be consistent with the parameters used in the seismic design and in the fragility analysis for the seismic probabilistic safety assessment, and it should be confirmed that the filter is set to capture all events with potential radiological consequences.

(b) Lower bound magnitude: The selected lower bound magnitude should not exceed $M_W = 5.0$.

(c) As an alternative to the use of a magnitude measure such as M_W, the lower bound motion filter may be specified in terms of an indicator of damage potential, such as cumulative absolute velocity, in conjunction with a specific value of that parameter for which it can be clearly demonstrated that no contribution to damage or risk will occur.

INDEPENDENT PEER REVIEW

10.15. In view of the complexity of the seismic hazard assessment, an independent peer review should be included as part of the project work plan and should be conducted to provide assurance that (a) a proper process has been duly followed in conducting the seismic hazard analysis, (b) the analysis has addressed and evaluated the uncertainties involved (both epistemic and aleatory), and (c) the documentation is complete and traceable.

10.16. Two methods of peer review should be used: participatory peer review and late stage peer review. A participatory peer review is carried out during the assessment, allowing the reviewers to resolve comments when technical issues arise in the process of the seismic hazard analysis. A late stage (follow-up) peer review is carried out towards the end of the assessment. Participatory peer review will decrease the likelihood that the assessment will be found unsuitable at a late stage.

10.17. The independent peer review should address all parts of the seismic hazard assessment, including the collection and evaluation of the available data, the process for the seismic hazard analysis, all technical elements (e.g. seismic source characterization, ground motion evaluation), the method of seismic hazard analysis, quantification of uncertainties and documentation. The procedure should be based on the participation of a duly qualified, multidisciplinary team of experts and the integration of their different professional judgements. The procedure should include the conduct of technical meetings or workshops to discuss the reliability and quality of available data, the safety significance of hazards, and alternative interpretations of these, as well as to provide feedback to the project team. The number and timing of these workshops should be established in the proposed work plan in accordance with the necessary level of effort and the available resources. The meetings should be duly documented and reported.

10.18. The independent peer review team members should include multidisciplinary experts to address all technical and process related aspects of the analysis. The peer reviewers should not have been involved in the development of the seismic hazard analysis and should not have a vested interest in the outcome. The level and type of peer review can differ, depending on the intended application of the seismic hazard analysis.

10.19. In dealing with issues relating to seismic hazard assessment, it may be possible for the project team to recognize and represent the centre, the body and the range of technically defensible interpretations through interactions with experts not directly involved with the project ('invited experts') who participate to provide their specific interpretation and professional judgement on the subject or issue under discussion. Such invited experts should provide their input to the independent peer review team, although they are not directly involved in the peer review. This approach is most suitable for topics that pertain to regional modelling issues; for issues pertaining to the near regional and the site vicinity scales, invited experts might not adequately provide diversity because they do not possess project specific data.

REFERENCES

[1] INTERNATIONAL ATOMIC ENERGY AGENCY, Site Evaluation for Nuclear Installations, IAEA Safety Standards Series No. SSR-1, IAEA, Vienna (2019).

[2]   INTERNATIONAL ATOMIC ENERGY AGENCY, IAEA Safety Glossary: Terminology Used in Nuclear Safety and Radiation Protection, 2018 Edition, IAEA, Vienna (2019).

[3]   INTERNATIONAL ATOMIC ENERGY AGENCY, Geotechnical Aspects of Site Evaluation and Foundations for Nuclear Power Plants, IAEA Safety Standards Series No. NS-G-3.6, IAEA, Vienna (2004).

[4]   INTERNATIONAL ATOMIC ENERGY AGENCY, WORLD METEOROLOGICAL ORGANIZATION, Meteorological and Hydrological Hazards in Site Evaluation for Nuclear Installations, IAEA Safety Standards Series No. SSG-18, IAEA, Vienna (2011).

[5]   INTERNATIONAL ATOMIC ENERGY AGENCY, Seismic Design for Nuclear Installations, IAEA Safety Standards Series No. SSG-67, IAEA, Vienna (2021).

[6]   NUCLEAR REGULATORY COMMISSION, Updated Implementation Guidelines for SSHAC Hazard Studies, Rep. NUREG-2213, Office of Nuclear Regulatory Research, Washington, DC (2018).

[7]   INTERNATIONAL ATOMIC ENERGY AGENCY, Ground Motion Simulation Based on Fault Rupture Modelling for Seismic Hazard Assessment in Site Evaluation for Nuclear Installations, Safety Reports Series No. 85, IAEA, Vienna (2015).

[8]   INTERNATIONAL ATOMIC ENERGY AGENCY, Diffuse Seismicity in Seismic Hazard Assessment for Site Evaluation of Nuclear Installations, Safety Reports Series No. 89, IAEA, Vienna (2016).

[9]   INTERNATIONAL ATOMIC ENERGY AGENCY, Safety of Nuclear Power Plants: Design, IAEA Safety Standards Series No. SSR-2/1 (Rev. 1), IAEA, Vienna (2016).

[10]   INTERNATIONAL ATOMIC ENERGY AGENCY, Safety of Research Reactors, IAEA Safety Standards Series No. SSR-3, IAEA, Vienna (2016).

[11]   INTERNATIONAL ATOMIC ENERGY AGENCY, Safety of Nuclear Fuel Cycle Facilities, IAEA Safety Standards Series No. SSR-4, IAEA, Vienna (2017).

[12]   INTERNATIONAL ATOMIC ENERGY AGENCY, Development and Application of Level 1 Probabilistic Safety Assessment for Nuclear Power Plants, IAEA Safety Standards Series No. SSG-3, IAEA, Vienna (2010). (A revision of this publication is in preparation.)

[13]   INTERNATIONAL ATOMIC ENERGY AGENCY, Site Survey and Site Selection for Nuclear Installations, IAEA Safety Standards Series No. SSG-35, IAEA, Vienna (2015).

[14]   INTERNATIONAL ATOMIC ENERGY AGENCY, Evaluation of Seismic Safety for Existing Nuclear Installations, IAEA Safety Standards Series No. NS-G-2.13, IAEA, Vienna (2009). (A revision of this publication is in preparation.)

[15]   INTERNATIONAL ATOMIC ENERGY AGENCY, Volcanic Hazards in Site Evaluation for Nuclear Installations, IAEA Safety Standards Series No. SSG-21, IAEA, Vienna (2012).

[16]   INTERNATIONAL ATOMIC ENERGY AGENCY, Leadership and Management for Safety, IAEA Safety Standards Series No. GSR Part 2, IAEA, Vienna (2016).

[17]   INTERNATIONAL ATOMIC ENERGY AGENCY, Application of the Management System for Facilities and Activities, IAEA Safety Standards Series No. GS-G-3.1, IAEA, Vienna (2006).

Annex

TYPICAL OUTPUT OF PROBABILISTIC
SEISMIC HAZARD ANALYSES

Table A–1 shows the typical outputs obtained from probabilistic seismic hazard analyses, including the format in which these outputs are generally presented.

TABLE A–1. TYPICAL OUTPUT OF PROBABILISTIC SEISMIC HAZARD ANALYSES

Output	Description	Format
Mean hazard curves	Mean hazard curves are calculated from the suite of hazard curves (i.e. the annual frequency of exceedance as a function of a ground motion parameter of interest) generated for the individual logic tree branches in the probabilistic seismic hazard analysis.	Mean hazard curves are generally reported for each ground motion parameter of interest in tabular and graphic format.
Fractile hazard curves	Fractile hazard curves represent certain fractiles of the distribution of the suite of hazard curves generated for the individual logic tree branches in the probabilistic seismic hazard analysis, on the assumption of a Gaussian distribution.	Fractile hazard curves are generally reported for each ground motion parameter of interest in tabular and graphic format. Unless otherwise specified in the work plan, fractile levels of 0.05, 0.16, 0.50, 0.84 and 0.95 are generally reported.
Uniform hazard response spectra	A uniform hazard response spectrum represents the spectral response values that have an equal annual frequency of exceedance, derived from the seismic hazard curves of individual frequencies or periods.	Mean and fractile uniform hazard response spectra are generally reported in tabular and graphic format. Unless otherwise specified in the work plan, the uniform hazard response spectra are generally reported for annual frequencies of exceedance of 10^{-2}, 10^{-3}, 10^{-4}, 10^{-5}, 10^{-6} or 10^{-7} and for fractile levels of 0.05, 0.16, 0.50, 0.84 and 0.95.

TABLE A–1. TYPICAL OUTPUT OF PROBABILISTIC SEISMIC HAZARD
ANALYSES (cont.)

Output	Description	Format
Magnitude–distance (M–D) deaggregation	The M–D deaggregation quantifies the relative contributions from earthquakes of different sizes and at different distances to the seismic hazard at a site.	The results of M–D deaggregation provide the relative contributions of earthquakes that occur in specified M–D ranges (i.e. 'bins') to the level of the ground motion parameter corresponding to a certain annual frequency of exceedance of interest on the estimated hazard curve. In general, the deaggregation is performed for the mean hazard and for the annual frequencies of exceedance to be used in the design or seismic probabilistic safety assessment, and is generally reported in tabular and graphic format (e.g. a three dimensional bar plot).
Mean and modal magnitudes and distances	From the results of an M–D deaggregation, the mean and modal magnitudes and distances of earthquakes that contribute to the hazard can be determined.	The mean and modal magnitudes and distances are generally reported for each ground motion parameter and level for which the M–D deaggregated hazard results are given. Unless otherwise specified in the work plan, these results are generally reported for response spectral frequencies of 1, 2.5, 5, 10 and 25 Hz and for peak ground acceleration.
Seismic source deaggregation	The seismic hazard at a site is a sum of the hazard from individual seismic sources modelled in the probabilistic seismic hazard analysis. A deaggregation on the basis of seismic sources provides an insight into the possible location and type of future earthquake occurrences.	The seismic source deaggregation is generally reported for the levels of ground motion parameter considered in the probabilistic seismic hazard analysis. The deaggregation is generally performed for the mean hazard and presented as a series of seismic hazard curves.

TABLE A–1. TYPICAL OUTPUT OF PROBABILISTIC SEISMIC HAZARD
ANALYSES (cont.)

Output	Description	Format
Ground motion time histories	Time histories are shown as waveforms representing the ground motion (e.g. acceleration) as a function of time. For the purposes of engineering analysis, time histories may be needed that are consistent with the results of the probabilistic seismic hazard analysis. The criteria for selecting and/or generating a time history may be specified in the work plan. Example criteria include the selection of time histories that are consistent with the mean and modal magnitudes and distances for a specified ground motion or annual frequency of exceedance.	The format for presenting ground motion time histories (e.g. acceleration unit, time interval) will generally be defined in the work plan (e.g. acceleration would be represented in terms of g at a time interval of 0.01 s).

CONTRIBUTORS TO DRAFTING AND REVIEW

Ake, J.	Nuclear Regulatory Commission, United States of America
Altınyollar, A.	International Atomic Energy Agency
Asfaw, K.E.	International Atomic Energy Agency
Baize, S.	Institute for Radiological Protection and Nuclear Safety, France
Coman, O.	International Atomic Energy Agency
Contri, P.	International Atomic Energy Agency
Dalguer, L.	Consultant, Switzerland
Ford, P.	Consultant, United Kingdom
Fukushima, Y.	International Atomic Energy Agency
Godoy, A.	Consultant, Argentina
Guerrieri, L.	National Institute for Environmental Protection and Research, Italy
Gürpınar, A.	Consultant, Turkey
Hok, S.	Institute for Radiological Protection and Nuclear Safety, France
Johnson, J.	Consultant, United States of America
Kalinkin, I.	JSC Atomenergoproekt, Russian Federation
Kammerer, A.	Consultant, United States of America
Labbé, P.	École spéciale des travaux publics, France
Morita, S.	International Atomic Energy Agency
Nakajima, M.	Central Research Institute of Electric Power Industry, Japan
Nishizaki, S.	International Atomic Energy Agency

Ono, M.	International Atomic Energy Agency
Renault, P.	swissnuclear, Switzerland
Secanell, R.	Fugro, Spain
Serva, L.	Consultant, Italy
Sugaya, K.	Nuclear Regulation Authority, Japan
Suzuki, A.	International Atomic Energy Agency
Tajima, R.	Nuclear Regulation Authority, Japan
Toro, G.	Lettis Consultants International, Inc., United States of America
Viallet, E.	Électricité de France, France
Wattelle, E.	Institute for Radiological Protection and Nuclear Safety, France
Weidenbrück, K.	Federal Ministry for the Environment, Nature Conservation and Nuclear Safety, Germany
Wu, C.	Nuclear Regulation Authority, Japan

IAEA
International Atomic Energy Agency

ORDERING LOCALLY

IAEA priced publications may be purchased from the sources listed below or from major local booksellers.

Orders for unpriced publications should be made directly to the IAEA. The contact details are given at the end of this list.

NORTH AMERICA

Bernan / Rowman & Littlefield
15250 NBN Way, Blue Ridge Summit, PA 17214, USA
Telephone: +1 800 462 6420 • Fax: +1 800 338 4550
Email: orders@rowman.com • Web site: www.rowman.com/bernan

REST OF WORLD

Please contact your preferred local supplier, or our lead distributor:

Eurospan Group
Gray's Inn House
127 Clerkenwell Road
London EC1R 5DB
United Kingdom

Trade orders and enquiries:
Telephone: +44 (0)176 760 4972 • Fax: +44 (0)176 760 1640
Email: eurospan@turpin-distribution.com

Individual orders:
www.eurospanbookstore.com/iaea

For further information:
Telephone: +44 (0)207 240 0856 • Fax: +44 (0)207 379 0609
Email: info@eurospangroup.com • Web site: www.eurospangroup.com

Orders for both priced and unpriced publications may be addressed directly to:
Marketing and Sales Unit
International Atomic Energy Agency
Vienna International Centre, PO Box 100, 1400 Vienna, Austria
Telephone: +43 1 2600 22529 or 22530 • Fax: +43 1 26007 22529
Email: sales.publications@iaea.org • Web site: www.iaea.org/publications